INDUSTRIAL PERSPECTIVE OF MICROBIOLOGY

A Sustainable Approach

Dr. Rachna Chaturvedi

Dr. Ruchi Yadav

Dr. Jyoti Prakash

INDIA · SINGAPORE · MALAYSIA

Copyright © Dr. Rachna Chaturvedi, Dr. Ruchi Yadav and Dr. Jyoti Prakash 2024
All Rights Reserved.

ISBN
Paperback 979-8-89610-643-2
Hardcase 979-8-89610-644-9

This book has been published with all efforts taken to make the material error-free after the consent of the author. However, the author and the publisher do not assume and hereby disclaim any liability to any party for any loss, damage, or disruption caused by errors or omissions, whether such errors or omissions result from negligence, accident, or any other cause.

While every effort has been made to avoid any mistake or omission, this publication is being sold on the condition and understanding that neither the author nor the publishers or printers would be liable in any manner to any person by reason of any mistake or omission in this publication or for any action taken or omitted to be taken or advice rendered or accepted on the basis of this work. For any defect in printing or binding the publishers will be liable only to replace the defective copy by another copy of this work then available.

TABLE OF CONTENTS

Foreword 7
Preface 9
Acknowledgments 11
Introduction 13

1. **INDUSTRIAL MICROBIOLOGY FOR SUSTAINABLE DEVELOPMENT: A FOCUS ON BIOFUELS AND FERMENTATION** 15
   Introduction 15
   Biofuels: a Sustainable Energy Source 17
   Fermentation Technology: the Backbone of Biofuel Production 18
   Integration of Biofuels and Fermentation Technology 20
       Feedstock Utilization 20
       Microbial Strain Optimization 20
       Process Integration 20
   Future Perspectives and Recommendations 21
   Conclusions 22

2. **MICROBIAL GROWTH DYNAMICS: UNRAVELLING THE IMPACT OF TEMPERATURE AND pH** 25
   Introduction 25
   Importance of Microbial Growth 25
   Temperature Effect on the Microbial Growth 26
   Ph Effect on the Microbial Growth 27
   Adaptation Mechanisms to Temperature and pH Extremes 28
   Influence of Temperature and pH on Enzyme Stability and Activity 28
   Role of Temperature and pH in Pathogen Virulence and Infectivity 30
   Industrial Applications of Temperature and pH Tolerant Microbes 30
   Antagonistic Effect on Microbial Growth 30

3. **BACTERIAL CONTRIBUTIONS TO ANTIBIOTIC
   DEVELOPMENT AND PRODUCTION** 33
   Introduction 33
     Natural Producers of Antibiotics 34
     Biotechnological Production 34
     Antibiotic Resistance Research 34
     Symbiotic Relationships 35
     Antibiotic Producing Bacteria 35
   Antibiotics Produced By Actinomycetes 36
   Streptomycin 36
     Mechanism of Action: 37
   Gentamycin 38
     Mechanism of Action 38
     Spectrum of Activity 38
     Biosynthesis Pathway 38
   Erythromycin 38
     Role of Bacteria in Erythromycin Production 39
   Bacitracin 39
     Biosynthesis Pathway 39
     Industrial Production 39
   B-polymixin 40
     Mechanism of Action: 40
     Spectrum Activity 40
     Methods for Production of Antibiotics 40

4. **LICHEN AS BIOINDICATORS FOR AIR POLLUTION DETECTION** 43
   Introduction 44
   Lichens 45
   Classification of Lichens 45
   Lichens Are Called as Pollution Indicator 46
   Major Lichen-sensitive Air Pollutants 46
   Lichens Are Used as Air Quality Indicator 47
   Circumference of at Least 60 Cm,no or Very Little Shade 47
   All Lichens Indicate That the Air is Clean 47
   Lichen-based Nitrogen Air Quality Index Guide 47

5. **IMPORTANCE OF ACETIC ACID BACTERIA IN FOOD INDUSTRY** 51
   Introduction 51
   Acetic Acid Bacteria and Alcoholic Beverages 52
     Lambic Beer 52
     Water Kefir 52
     Kombucha 53
   Acetic Acid Bacteria and Vinegar Production 54
   Cocoa and Acetic Acid Bacteria 56
   Nata De Coco and Acetic Acid Bacteria 56
   Nata De Piña and Acetic Acid Bacteria 57

Palm Wine Vinegar and Acetic Acid Bacteria 57
Pickles and Acetic Acid Bacteria. 57
Future Prospects of Acetic Acid Bacteria 58

6. IMPORTANCE OF LACTIC ACID BACTERIA IN FOOD INDUSTRY 61
Introduction 61
Significance of Lab in the Food Industry 63
Lactic Acid Bacteria in Food Production 64
    Starter Culture 64
    Non-starter Cultures 64
    Bio-protective Cultures 65
Lab in Preserving and Ensuring Safety of Food 65
Lab Against Foodborne Bacterial Pathogens 66
Lab Against Yeast 66
Lab Against Filamentous Fungi 66

7. HARNESSING BACTERIA FOR SUSTAINABLE CROP IMPROVEMENT 69
Introduction 70
Crops and Bacterial Association 71
biotechnological Advancements in Bacterial Crop Improvement 71
Importance of Bacteria 72
Biological Nitrogen Fixation 73
Agricultural Forecasting 74
Environmental Conservation 75
Rural Development 76

8. ALGAL LIPID COMPONENT: A PROMISING TOOL IN BIOTECHNOLOGY ... 81
Introduction 81
Pharmaceutical 84
Cosmetics 85

9. MICROBIAL BIOTECHNOLOGY IN FOOD INDUSTRY 93
Introduction 94
Role of Microbes in Food Production 94
Microbial Enzymes in Food Processing: 95
Microbial Safety and Quality Control 96
Advances in Microbial Biotechnology 97

10. GOOD BACTERIA AND MICROBIOME FOR HUMAN HEALTH 101
Introduction 101
The Human Gastrointestinal Tract Microbiome 103
The Human Respiratory Tract Microbiome 104
The Human Reproductive Tract Microbiome 105
The Human Skin Microbiome 106

11. INTEGRATING BIOINFORMATICS INTO BIOREMEDIATION STRATEGIES ... 111
Introduction ... 111
Bioremediation ... 113
Proteomics in Bioremediation ... 113
Proteomics Tools and Techniques Applied in Bioremediation ... 113
Genomics in Bioremediation ... 114
Genomics Tools and Techniques Applied in Bioremediation ... 115
Phylogeny in Bioremediation ... 115
Bioinformatics Databases Applied in Bioremediation ... 116

12. MICROBIAL GENOMICS AND ITS INDUSTRIAL APPLICATIONS ... 119
Introduction ... 119
History and Development of Microbial Genomics ... 121
    Waste Management Methods ... 121
    Industrial Waste Composting ... 121
    Physicochemical Method ... 122
    Algae in Industrial Waste Management ... 122
Phytoremediation ... 122

13. GUT METAGENOMICS: A NEW AVENUE FOR DRUG DEVELOPMENT ... 125
Introduction to Gut Metagenomics ... 126
Historical Context and Advancements ... 126
Overview of the Human Gut Microbiome ... 126
Techniques in Gut Metagenomics ... 127
    Sampling and Sequencing Methods ... 127
Bioinformatics Tools for Data Analysis ... 128
Challenges and Limitations ... 129
Gut Metagenome Role in Digestion and Metabolism ... 129
Gut Metagenome and Immune System ... 130
Diseases and Disorders ... 130
Drug Discovery Through Gut Metagenomics ... 130
    Identification of Novel Drug Targets ... 130
Screening for New Bioactive Compounds ... 131
Case Study ... 132
Future Directions and Innovations in Gut Metagenomics ... 133

14. INDUSTRIAL WASTE MANAGEMENT BY USING MICROORGANISMS ... 137
Introduction ... 138
Types of Industrial Waste ... 138
Management of Industrial Waste ... 139
Waste Management Methods ... 140
Industrial Waste Composting ... 140
Physicochemical Method ... 140
Phyco Bioremediation ... 141

Epilogue ... 145

FOREWORD

Microbiology, the study of microscopic organisms, has long been a cornerstone of scientific discovery and innovation. From the initial observations of microorganisms by Antonie van Leeuwenhoek in the 17th century to the intricate genetic engineering feats of the 21st century, microbiology has continuously expanded our understanding of the living world. The implications of these advancements are profound, particularly within the industrial sector, where the application of microbiological principles has revolutionized numerous industries, from pharmaceuticals and food production to biotechnology and environmental management.

"Industrial Perspective of Microbiology" delves into this fascinating intersection of science and industry. This book aims to bridge the gap between fundamental microbiological research and its practical applications in various industrial contexts. It highlights how microorganisms, once perceived solely as agents of disease, are now recognized as pivotal players in sustainable industrial processes and biotechnological innovations.

The chapters within this book cover a broad spectrum of topics, each meticulously curated to provide a comprehensive understanding of the industrial applications of microbiology. Readers will explore the roles of microbes in fermentation processes, bioremediation, biofuel production, and the development of novel pharmaceuticals. The book also addresses the latest advancements in microbial genomics and synthetic biology, underscoring the cutting-edge techniques that are propelling the industry forward.

This book is not only a testament to the remarkable capabilities of microorganisms but also a celebration of the scientists and engineers who harness these capabilities for the betterment of society. Their work underscores a fundamental truth: the smallest living entities can have the most significant impact.

We are confident that the "Industrial Perspective of Microbiology" will serve as an invaluable resource for students, researchers, and professionals alike. It will inspire a deeper appreciation for the role of microbiology in industrial innovation and encourage further exploration into this dynamic and ever-evolving field.

As we continue to face global challenges, from environmental degradation to the need for sustainable energy sources, the insights and applications presented in this book will be crucial in guiding us toward innovative solutions. The future of industrial microbiology is bright, and this book is a vital contribution to that promising future

PREFACE

Microbiology, the study of microscopic organisms, has revolutionized numerous industries, reshaping the way we understand and interact with the microscopic world. From the fermentation processes utilized in brewing and baking to the development of antibiotics and biotechnology applications, the impact of microbiology on industrial processes is both profound and far-reaching.

This book, "Industrial Perspective of Microbiology," aims to bridge the gap between foundational microbiological concepts and their practical applications in various industries. It explores how microorganisms are harnessed to drive innovation, enhance productivity, and solve complex problems in sectors such as pharmaceuticals, agriculture, food production, and environmental management.

The journey through this book begins with an introduction to the fundamental principles of microbiology, providing a solid grounding for readers new to the field. Subsequent chapters delve into specific industrial applications, highlighting the pivotal role microorganisms play in each domain. Through detailed case studies, practical examples, and insights from leading experts, we illustrate the transformative potential of microbiology in industrial settings.

Our objective is to equip readers with a comprehensive understanding of the intersection between microbiology and industry. Whether you are a student, researcher, or professional, this book is designed to inspire and inform, offering valuable knowledge and practical tools to harness the power of microorganisms.

We extend our gratitude to the many contributors who have shared their expertise and experiences, enriched the content, and broadened the perspectives presented. Their collective insights form the backbone of this work, reflecting the dynamic and interdisciplinary nature of industrial microbiology.

As we navigate the challenges and opportunities of the 21st century, the role of microbiology in industry will continue to evolve and expand. We hope that this book serves as both a resource and a catalyst, fostering innovation and encouraging the sustainable application of microbiological principles in industrial contexts.

Thank you for embarking on this journey with us. We invite you to explore the fascinating world of industrial microbiology and discover the myriad ways microorganisms are shaping our world.

ACKNOWLEDGMENTS

Editing this book has been a journey that we could not have completed alone. We want to extend our deepest gratitude to all those who have supported me throughout this process.

With utmost veneration, we would like to join hands and express our humble gratitude to *God Almighty,* who bestowed upon us the necessary strength, courage, and good health for the completion of this book. Without His grace, it would not have been possible for us to bring accomplish this Herculean task.

I wish to express my acknowledgment to the ***Honorable Dr. Ashok K. Chauhan;*** Founder President, Ritnand Balved Education Foundation (RBEF); ***Dr. Aseem Chauhan;*** Chairman, Amity University, Uttar Pradesh, Lucknow Campus, ***Prof. (Dr.) Anil Vashisht*** Pro Vice-Chancellor, AUUP, Lucknow Campus **Wg.Cdr (Dr.) Anil Kumar Tiwari** Deputy Pro Vice-Chancellor, AUUP, Lucknow Campus; ***Dr. Janmejai Kumar Srivastava; Director***, Amity Institute of Biotechnology (AIB), Amity University Utter Pradesh (AUUP), for their constant support.

We owe a special thanks to our authors, who contributed their valuable chapters and made this book a great success and fulfilled the theme of this book in all concepts.

We are also very thankful to our family members for their belief, support, and encouragement throughout. We are also grateful to my friends and colleagues, thank you for your encouragement and for providing me with the much-needed balance between work and relaxation.

Finally, we are deeply grateful to my readers. Your interest and support are what make this journey worthwhile. We hope this book meets your expectations and provides you with valuable insights.

Thank you all for being part of this journey with us.

INTRODUCTION

"Industrial Perspective of Microbiology" is a comprehensive exploration of the pivotal role microbiology plays in various industrial applications. This book delves into the utilization of microorganisms to produce a wide array of products, including pharmaceuticals, enzymes, biofuels, and food additives. It highlights the advancements in biotechnology and genetic engineering that have expanded the capabilities of industrial microbiology, enabling the development of more efficient and sustainable processes.

Microbiology plays a crucial role in various industrial sectors, contributing to advancements in medicine, food production, environmental management, and biotechnology. Here's an overview of its industrial perspectives:

Key topics covered in the book include:
- Agriculture: Microbiology contributes to agricultural productivity through the development of microbial pesticides, biofertilizers, and plant growth-promoting microbes. It also involves research on soil microbiota to enhance soil health and fertility.
- Microbial Biotechnology: The use of microbes in the production of industrially important products. This section explores the genetic manipulation of microorganisms to enhance their productivity and functionality. Microorganisms serve as workhorses in biotechnological applications, including the production of biofuels, enzymes, and biopolymers. Genetic engineering techniques enable the manipulation of microbial genomes for the synthesis of valuable compounds and the development of novel bioproducts.
- Fermentation Technology: An in-depth look at the fermentation process, including the types of fermenters, optimization of fermentation conditions, and the scale-up process from laboratory to industrial scale.
- Bioprocess Engineering: The principles and applications of bioprocess engineering in the design and operation of industrial-scale microbial processes.
- Pharmaceutical Microbiology: The role of microorganisms in the production of antibiotics, vaccines, and other pharmaceuticals. This section also addresses the challenges and solutions to ensuring product safety and efficacy. Microbiology is integral to the pharmaceutical industry for the development of antibiotics, vaccines, and other drugs. Microorganisms are used in fermentation processes to produce antibiotics like penicillin and streptomycin. They are also employed in the production of recombinant proteins and vaccines, such as those for hepatitis B and human papillomavirus (HPV).
- Food Microbiology: The use of microorganisms in the production and preservation of food products. Topics include fermentation processes in the dairy, brewing, and baking industries. In the food industry, microbiology ensures food safety and quality. It involves the detection and control of pathogens, spoilage microorganisms, and foodborne toxins. Microorganisms

are also harnessed in food fermentation processes, such as those used in brewing, cheese-making, and yogurt production.
- Environmental Microbiology: The application of microbial processes in environmental management, including bioremediation, waste treatment, and the production of biofuels. Microbiology aids in ecological monitoring and remediation efforts. Microbes are utilized in wastewater treatment plants to break down organic pollutants and in bioremediation processes to clean up oil spills and contaminated soil.
- Diagnostic Testing: Microbiology is vital in clinical diagnostics to identify infectious agents responsible for diseases. Techniques like polymerase chain reaction (PCR), immunoassays, and microbial culture facilitate the detection and characterization of pathogens, aiding in disease diagnosis and treatment.
- Regulatory and Quality Control: An overview of the regulatory frameworks governing industrial microbiology and the importance of quality control in ensuring product consistency and safety.
- Cosmetics and Personal Care: Microbiology ensures the safety and stability of cosmetics and personal care products. It involves microbial testing to assess product quality and to prevent microbial contamination, which could lead to product spoilage or adverse effects on consumers.

Overall, microbiology provides the foundation for numerous industrial processes and innovations, driving advancements in healthcare, agriculture, environmental sustainability, and consumer products. Its interdisciplinary nature fosters collaboration between microbiologists, engineers, chemists, and other professionals to address global challenges and improve the quality of life.

The book serves as a valuable resource for students, researchers, and professionals in the fields of microbiology, biotechnology, and industrial engineering, providing a solid foundation in both the theoretical and practical aspects of industrial microbiology. Through detailed case studies and real-world examples, readers gain insights into the innovative applications of microorganisms in various industries and the future trends shaping this dynamic field.

CHAPTER 1

INDUSTRIAL MICROBIOLOGY FOR SUSTAINABLE DEVELOPMENT: A FOCUS ON BIOFUELS AND FERMENTATION

Siddharth Singh, Jyoti Prakash, and Rachna Chaturvedi

Amity Institute of Biotechnology, Amity University Uttar Pradesh, Lucknow, 226028

Abstract

Industrial microbiology holds immense potential to drive sustainable development by providing eco-friendly alternatives to conventional processes. This paper explores the role of industrial microbiology in promoting sustainable development, particularly through biofuel production and fermentation technology. Biofuels produced through microbial fermentation of renewable resources offer a promising avenue to reduce greenhouse gas emissions and mitigate climate change. Fermentation technology, the backbone of biofuel production, uses microorganism metabolic activities to convert biomass into biofuels like ethanol and biodiesel. By integrating these technologies, we can reduce our reliance on fossil fuels, promote a circular economy, and decrease global warming. By optimizing microbial strains, fermentation processes, and downstream recovery, significant advancements can be achieved to enhance biofuel yields, reduce production costs, and minimize environmental impact. This abstract underscore the importance of integrating industrial microbiology with sustainable practices to create a circular economy and promote a greener future.

Keywords:
sustainable development, biofuels, fermentation technology, renewable resources.

INTRODUCTION

Microorganisms play a big role in the industry, and they can be used in multiple ways. Microbes can be employed medicinally to produce antibiotics to cure infections. The food business can potentially benefit from the usage of microbes. Industrial microbiology is also a fascinating field that utilizes microorganisms to create valuable products on a large scale. It bridges the gap between microbiological research and practical applications, focusing on the utilization of bacteria, fungi, yeasts, and other microbes to improve industrial processes, create novel products, and address environmental challenges. Meaning it is a discipline that applies microbial and fermentation processes to develop products, processes, and services that can improve the quality of life and environmental sustainability.

Historically, industrial microbiology has played a crucial role in the production of antibiotics, vitamins, and other essential compounds, revolutionizing fields such as medicine and agriculture. Today, its scope has expanded to include applications in biotechnology, environmental management, and sustainable development. Microorganisms are employed to produce biofuels, bioplastics, and specialty chemicals, offering alternatives to traditional petrochemical-based products. Our current practices, including the indiscriminate use of chemicals, increased employment of non-renewable sources of energy, and uncontrolled generation of waste products in every possible industrial process, have posed a large threat to the sustainability of the environment. The world is currently under more pressure to implement green technologies, cleaner manufacturing practices, and sustainable policies to preserve Earth's biosphere for coming generations [1]. Sustainable development is greatly aided by microbial technology in a variety of sectors and applications. By harnessing the power of microorganisms, we can optimize waste management, reduce greenhouse gas emissions, and produce renewable energy sources, this contributes significantly to sustainable development goals (SDGs). Addressing significant global challenges related to resource depletion and environmental sustainability, such as food security, healthcare provision, well-being improvement, and green energy development, becomes possible through these efforts. Providing knowledge about the beneficial roles that microbes play in society might encourage the wise use of microorganisms in sustainable development [2].

The primary factor promoting the bioremediation of organic and inorganic materials at the contaminated sites is microbial diversity [3]. The multiphasic distribution of pollutants due to increased human activities makes the bioremediation process more complex. Through metabolic activity, the bacteria can convert complicated contaminants into simpler, biodegradable molecules. Nevertheless, the combination of contaminants hinders microorganisms' capacity to biodegrade [4]. Process parameters and microorganisms can be optimized to create effective decontamination solutions. Global initiatives for sustainable development combined with the rising energy demands of urbanization and industrialization have led to a greater emphasis on renewable energy projects [5]. Green microbiology focuses on harnessing the power of microorganisms to provide eco-friendly solutions and contribute to environmental sustainability. Furthermore, it aims to reduce industries' environmental footprint by incorporating microbial technologies in food production, energy generation, waste management, and bioremediation. The objective is to encourage environmentally friendly behaviors by utilizing microorganisms in business and providing eco-friendly solutions for various industries. In recent times, scientists have been working to understand and improve naturally available microbial processes for environmental sustainability. The literature on industrial microbiology's contributions to sustainable development has grown significantly over the past few decades. Key areas of focus include the development of microbial strains capable of converting biomass into biofuels, the exploration of fermentation technology for waste valorization, and the integration of these processes into existing industrial frameworks.

Kafarski developed a color coding system to differentiate the main areas of biotechnology: white (industrial), green (agricultural), blue (marine and freshwater), red (pharmaceutical), brown (desert biotechnology), and purple (patents and inventions), among others. With the help of industrial-scale manufacturing of enzymes that can break down plant polymers, he explains how white biotechnology has the potential to completely transform the energy industry by producing liquid and gaseous biofuels from waste organic substrates sustainably [6]. Several studies have emphasized the feasibility of using microorganisms for biofuel production, showcasing the potential of bacteria and yeast in fermenting sugars derived from agricultural residues. [7]. By combining engineering-based technologies with selection, Karyolaimos and colleagues describe adjustments to E. coli strains for the generation of prokaryotic and eukaryotic recombinant membrane proteins. This involves adjusting the synthesis of membrane proteins to the machinery of membrane biogenesis [8]. For a very long time, lactic acid bacteria (LAB) have been used in industrial processes that include food starter cultures, and the synthesis of various metabolites like lactic acid, favor-enhancers, and intermediates for the plastics and pharmaceutical sectors. However, the complicated nutritional needs of LAB sometimes restrict their applicability for utilization in financially feasible biorefinery processes. Tarraran and Mazzoli address the traits of LAB strains that were isolated from plant settings where the primary source of carbon is plant material. Though they are not cellulolytic microbes on their own, these LAB can ferment a wide range of sugars generated from lignocellulosic biomass. The articles discuss how different biomass pre-treatments and co-cultivation methods have been created to help produce lactic acid at yields that are nearly the theoretical maximum [9]. Novak and Pflügl discuss the importance of acetate, obtained from plant and food waste and gaseous substrates such as CO, CO_2, and H_2, as a substrate to increase the performance of microbial cell factories [10]. Subsequent research has revealed that various strains could produce bioethanol, biodiesel, and biogas, establishing a direct connection between microbial processes and renewable energy generation [11]. Furthermore, advancements in genetic engineering have enabled the development of tailored microbial strains with enhanced metabolic pathways, increasing the efficiency of biofuel production [12]. Fermentation technology has also progressed rapidly, facilitating the large-scale conversion of waste materials into valuable biofuels and other bioproducts, thereby promoting a circular economy [13]. In addition to biofuel production, industrial microbiology is increasingly recognized for its role in reducing environmental impact through bioremediation and waste treatment. Microorganisms are used to degrade pollutants in soil and water, highlighting their invaluable capacity to restore ecological balance. This body of literature underscores the multifaceted contributions of industrial microbiology to sustainable development.

BIOFUELS: A SUSTAINABLE ENERGY SOURCE

Biofuels are renewable energy sources derived from biological materials, which can serve as alternatives to fossil fuels. Climate change, air pollution, and energy security concerns also necessitate a shift towards renewable energy sources. The main categories of biofuels include bioethanol, biodiesel, and biogas, all of which can be produced through microbial fermentation processes. Unlike fossil fuels, biofuels are often carbon-neutral, as the carbon dioxide released during combustion is offset by the carbon absorbed during the growth of the biomass. A good alternative fuel that comes almost exclusively from food crops is bioethanol. Whereas biodiesel because of its advantages for the environment, has grown in popularity recently and for all these productions, biofuels heavily depend on Industrial microbiology. Microorganisms such as *Saccharomyces cerevisiae* (yeast) are employed in the fermentation of sugars extracted from crops like corn, sugarcane, or cellulose-rich waste materials. Furthermore, bacteria such as *Clostridium* species can convert lignocellulosic biomass into biofuels through various anaerobic fermentation processes [12]. The versatility of industrial microbiology allows for the utilization of a wide range of feedstocks, making biofuels a viable solution for reducing dependence on fossil fuels. Biofuels are broadly classified into three generations, each with distinct characteristics and production methods. First-generation biofuels are produced from edible plant materials, while second-generation biofuels are derived from non-edible plant materials and agricultural residues and Third-generation biofuels are derived from microorganisms like algae and offer higher sustainability benefits due to their high yields and minimal impact on land and water resources.

Table 1: Classification of Biofuels

Generation	Feedstock	Example
I	corn, sugarcane, vegetable oils, food wastes	Ethanol, Biodiesel (B100; B20), Biogas
II	Cellulose, wood, straw, switchgrass, algae, non-food waste	Bio-butanol, Cellulosic ethanol, Algae biofuels
III	Algae	Algal biofuels

Comparing biofuels to fossil fuels has several advantages for the environment and the economy. They reduce greenhouse gas emissions, produce renewable resources, and reduce air pollution. They can also be produced from agricultural and industrial waste, reducing waste disposal and recycling. Biofuels also provide energy security by diversifying energy supply, reducing dependence on imported fossil fuels, and stimulating rural economies by creating jobs in agriculture and processing industries. Additionally, biofuels drive technological innovation, as they drive advancements in agriculture, biotechnology, and renewable energy technologies.

FERMENTATION TECHNOLOGY: THE BACKBONE OF BIOFUEL PRODUCTION

Fermentation technology is a critical process in biofuel production, serving as the backbone of converting organic materials into valuable energy resources. Fermentation is a metabolic process where microorganisms convert carbohydrates (such as sugars) into alcohols or acids, releasing energy in the form of ATP and producing byproducts like ethanol or butanol. This process is central to the production of biofuels, particularly ethanol and butanol, which are used as renewable alternatives to fossil fuels. The fermentation process involves several stages, including feedstock preparation, saccharification, fermentation, distillation and purification, and waste management. To make feedstocks more microbially friendly, they are ground, milled, and treated beforehand. Saccharification is necessary for complex carbohydrates like starch or cellulose, breaking them down into simpler sugars using enzymes or acid hydrolysis. The sugars are introduced to a fermentation vessel, where microorganisms like yeast or bacteria metabolize them to produce desired biofuels. Environmental factors like temperature, pH, and oxygen levels are controlled to optimize microbial activity. After fermentation, the product mixture contains ethanol, byproducts, and residual biomass, which are separated, purified, and removed to produce fuel-grade ethanol. Waste management involves managing byproducts, such as residual biomass and spent yeast, for animal feed, compost, or other uses.

Industrial Microbiology for Sustainable Development: A Focus on Biofuels and Fermentation

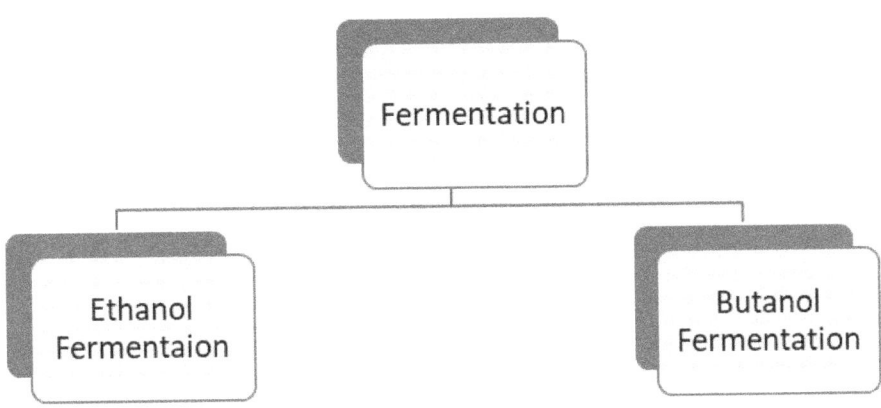

Figure 1: Types of fermentation for biofuel production

Fermentation technology offers several advantages in biofuel production, including renewable biomass sources, carbon neutrality, economic viability, and technological innovation. Renewable biomass sources enable a sustainable and circular energy economy, unlike finite fossil fuels. Fermentation also minimizes net carbon emissions, making biofuels carbon-neutral. The widespread availability of biomass feedstocks and advancements in fermentation techniques make biofuel production cost-effective, enhancing energy security and independence. Technological innovation, such as genetically engineered microorganisms and optimized fermentation processes, further enhances efficiency and yield, reducing production costs and improving sustainability. Recent advancements in fermentation technology aim to improve efficiency, reduce costs, and expand feedstocks and products. Genetic engineering involves genetic modifications of microorganisms to ferment a wider range of substrates and increase yield [15]. Enzyme technologies, such as cellulases and amylases, have enhanced saccharification processes, making them more economical and effective. Integrated biorefineries integrate multiple processes, including fermentation, to maximize resource utilization and output, converting biomass into a spectrum of biofuels and byproducts, optimizing efficiency and sustainability [16].

Table 2: List of microorganisms producing biofuels

Biofuel	Microorganisms	Yield (g/L)	References
Ethanol	Saccharomyces cerevisiae	50	[16]
	Zymomonas mobilis	77	[17]
Butanol	Escherichia coli	6.2	[18]
	Clostridium acetobutylicum	19.1	[16]
Hydrogen	Rhodopseudomonas palustris	3-8	[19]
Biodiesel	Chlamydomonas reinhardtii	12-25	[20]
Hydrocarbons	Botryococcus braunii	30-40	[21]
Fatty acids	Corynebacterium glutamicum	50	[22]
	Saccharomyces cerevisiae	0.38	[23]

INTEGRATION OF BIOFUELS AND FERMENTATION TECHNOLOGY

The integration of biofuels and fermentation technology is a dynamic and evolving field aimed at optimizing the production of renewable energy sources. By aligning the processes of biofuel production with advanced fermentation techniques, significant strides can be made in efficiency, cost-effectiveness, and sustainability. There are many ways by which we can integrate biofuel and ferment by focusing on feedstock utilization, microbial strain optimization, process integration, and the synergistic benefits arising from their combination.

Feedstock Utilization

Feedstock Selection and Preparation Effective integration of biofuels and fermentation technology begins with the selection and preparation of feedstocks. Feedstocks can be generically divided into three categories: first-generation, second-generation, and third-generation. Each category has its advantages and disadvantages. First-generation feedstocks include food crops like corn, sugarcane, and wheat, which are used for ethanol production in the US. Second-generation feedstocks include lignocellulosic biomass like agricultural residues and energy crops, which require pretreatment and saccharification to break down complex cellulose into fermentable sugars [24]. Third-generation feedstocks, such as algae, are promising due to their high yield of lipids and carbohydrates, potential for biodiesel or ethanol production, and their potential for non-arable land cultivation using wastewater [25].

Pretreatment and Saccharification The pretreatment of feedstocks is critical for breaking down complex carbohydrates into simpler sugars that can be fermented. Techniques include chemical (such as acid or alkaline), biological (such as enzyme), and physical (such as milling) treatments. Physical pretreatment involves grinding and milling to increase feedstock surface area [26], while chemical pretreatment involves acid or alkaline treatments to dissolve hemicellulose and remove lignin, and biological pretreatment involves enzymatic hydrolysis using cellulases to break down cellulose into glucose [27].

Microbial Strain Optimization

- **Yeast Strains for Ethanol Production** Genetic engineering has improved yeast strains, primarily Saccharomyces cerevisiae, for ethanol fermentation, enhancing their tolerance to ethanol and ability to utilize diverse feedstocks. These strains can ferment xylose and glucose, enhancing their ability to produce high-quality ethanol [14].
- **Bacterial Strains for Butanol Production** Clostridium acetobutylicum, a strain used for butanol production through acetone-butanol-ethanol fermentation, has been enhanced through metabolic engineering, enhancing its ability to ferment non-traditional substrates and increasing butanol tolerance [28].
- **Advanced Microbial Technologies** Synthetic biology and CRISPR technology are revolutionizing biofuel production by creating custom-built microorganisms with enhanced metabolic pathways and allowing precise genetic modifications, thereby improving strain performance and stability. [29][30].

Process Integration

- **Reactor Design:** Fermentation processes can be integrated by optimizing reactor design to improve efficiency and scalability. Types include batch, fed-batch, and continuous systems. Batch reactors are simple but not ideal for large-scale production due to product recovery inefficiencies. Fed-batch reactors allow controlled nutrient addition, improving yields and preventing substrate inhibition. Continuous reactors offer steady-state operation, enhancing productivity and efficiency for biofuel production [31].

- **Process Conditions:** Optimal conditions like temperature, pH, and oxygen levels are crucial for maximizing microbial performance and biofuel yield. Temperature control is essential for enzyme activity and growth [32], while pH control is crucial for efficient fermentation [33]. Oxygen levels are essential for aerobic fermentation, while anaerobic conditions are required for other types of fermentation.
- **Downstream Processing:** Efficient downstream processing is crucial for separating and purifying biofuels from fermentation broths. Techniques like distillation, filtration, and centrifugation are used to separate ethanol from water and other byproducts, while filtration and centrifugation remove solids and concentrate the desired biofuel [34][35].
- **Synergistic Benefits:** Integrating biofuels and fermentation technologies improves efficiency and reduces costs by optimizing feedstock processing, microbial performance, and reactor design. This leads to increased yield and productivity in biofuel production. Advances in enzyme technologies and reactor designs also contribute to cost reduction [36]. The combination of biofuels and fermentation technologies supports sustainability goals by utilizing renewable resources, reducing greenhouse gas emissions, and enhancing waste management and resource utilization [37].

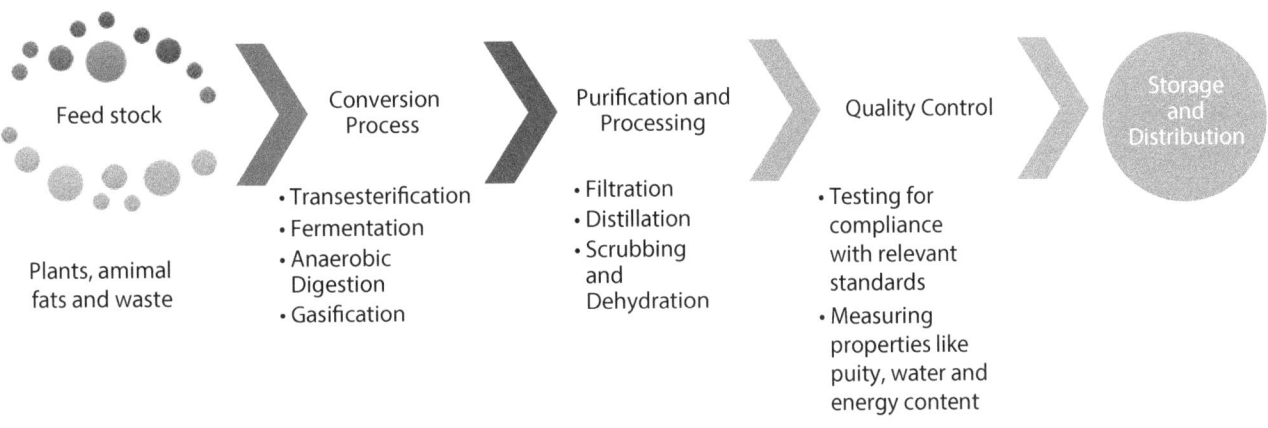

Figure 2: General flow diagram of Biofuel production

FUTURE PERSPECTIVES AND RECOMMENDATIONS

The future of biofuels will depend on addressing current challenges and capitalizing on their benefits. Key areas for future development include technological advancements, sustainable practices, policy support, and public acceptance. Technological advancements in genetic engineering, enzyme technologies, and process optimization are crucial for improving biofuel production efficiency and cost-effectiveness. Sustainable practices, such as using marginal lands for biofuel crops and adopting conservation tillage, can mitigate the environmental impact of biofuel production. Government policies and incentives, such as subsidies, tax credits, and research funding, can encourage investment and development in the biofuel sector.

However, challenges persist in fermentation technology, such as cost reduction, scalability, and sustainability. Research into more efficient microorganisms and cheaper substrates is crucial for reducing costs. Scaling up fermentation processes requires innovations in reactor design and process engineering. Developing more sustainable practices and integrating waste-to-energy technologies is also essential [38]. Advancements in microbial engineering, such as synthetic biology and gene editing technologies, hold promise for enhancing microbial strains used in fermentation. Research into new feedstocks, including waste materials and genetically engineered crops, aims to expand feedstock options and improve sustainability. Integrated biorefineries, which produce multiple biofuels and byproducts from a single feedstock, represent a significant step forward in optimizing biofuel production systems [39].

CONCLUSIONS

Biofuels are a promising and sustainable energy source that can reduce our reliance on fossil fuels and mitigate climate change impacts. They offer environmental and economic benefits, such as reduced greenhouse gas emissions, enhanced energy security, and rural development. However, challenges like land use change, water consumption, and production costs must be addressed. Fermentation technology is crucial in the biofuel industry, providing a sustainable method for producing ethanol, butanol, and other biofuels. Advancements in microbial engineering, enzyme technology, and process optimization address cost, scalability, and sustainability challenges. The integration of biofuels and fermentation technology is a crucial advancement in the pursuit of sustainable energy solutions. By harmonizing feedstock preparation, microbial optimization, and process technologies, the biofuel industry can achieve greater efficiency, cost-effectiveness, and environmental sustainability. Despite challenges, ongoing research and technological innovations hold promise for addressing these issues and advancing the field.

REFERENCES:

1. *Akinsemolu, A.A. (2018). The role of microorganisms in achieving the sustainable development goals. Journal of Cleaner Production, 182, 139-155.*
2. Kim, J. W., Park, S. B., Tran, Q. G., Cho, D. H., Choi, D. Y., Lee, Y. J., & Kim, H. S. (2020). Functional expression of polyethylene terephthalate-degrading enzyme (PETase) in green microalgae. *Microbial cell factories*, 19(1), 97.
3. Mali, H., Shah, C., Raghunandan, B. H., Prajapati, A. S., Patel, D. H., and Trivedi, U. (2022). Organophosphate pesticides an emerging environmental contaminant: pollution, toxicity, bioremediation progress, and remaining challenges. J. Environ. Sci. 127, 234–250.
4. Singh, A., & Mishra, V. K. (2021). Biodegradation of organic pollutants for its effective remediation from the environment and the role of various factors affecting the biodegradation process. In *Sustainable Environmental Clean-up* (pp. 1-27). Elsevier.
5. Kiliç, M., and Özdemir, E. (2018). "Chapter 1.7—Long-term energy demand and supply projections and evaluations for Turkey," in Exergetic, Energetic, and Environmental Dimensions, eds I. Dincer, C. O. Colpan, and O. Kizilkan (Cham: Elsevier Publishing), 115–132.
6. Kafarski, Pawel. (2012). Rainbow code of biotechnology. Chemik. 66. 814-816.
7. Manswama Boro, Ashwani Kumar Verma, Dixita Chettri, Vinod Kumar Yata, Anil Kumar Verma, Strategies involved in biofuel production from agro-based lignocellulose biomass, Environmental Technology & Innovation, Volume 28, 2022, 102679, ISSN 2352-1864

8. Karyolaimos A, Ampah-Korsah H, Zhang Z, de Gier JW. Shaping Escherichia coli for recombinant membrane protein production. FEMS Microbiol Lett. 2018 Aug 1;365(15). doi: 10.1093/femsle/fny152. PMID: 30007322.
9. Tarraran L, Mazzoli R. Alternative strategies for lignocellulose fermentation through lactic acid bacteria: the state of the art and perspectives. FEMS Microbiol Lett. 2018 Aug 1;365(15). doi: 10.1093/femsle/fny126. PMID: 30007320.
10. Novak K, Pflügl S. Towards biobased industry: acetate as a promising feedstock to enhance the potential of microbial cell factories. FEMS Microbiol Lett. 2018 Oct 1;365(20). doi: 10.1093/femsle/fny226. PMID: 30239700.
11. Tshikovhi A, Motaung TE. Technologies and Innovations for Biomass Energy Production. *Sustainability*. 2023; 15(16):12121.
12. Rathore D, Sevda S, Prasad S, Venkatramanan V, Chandel AK, Kataki R, Bhadra S, Channashettar V, Bora N, Singh A. Bioengineering to Accelerate Biodiesel Production for a Sustainable Biorefinery. *Bioengineering*. 2022; 9(11):618.
13. Rogers, T., & Brand, C. (2019). Advances in fermentation technology for biofuel production. *Bioresource Technology Reports*, 6, 1-10.
14. Zhao, X., & Jiao, J. (2015). *Genetic engineering strategies for enhancing microbial fermentation. Journal of Biotechnology, 213*, 29-35.
15. Morris, D. D., & Arora, A. (2016). *Design and operation of integrated biorefineries for advanced biofuels. Renewable Energy, 87*, 248-257.
16. Xue, C., Zhao, J., Lu, C., Yang, S. T., Bai, F., & Tang, I. C. (2012). High-titer n-butanol production by clostridium acetobutylicum JB200 in fed-batch fermentation with intermittent gas stripping.
17. Charoenpunthuwong K, Klanrit P, Chamnipa N, Thanonkeo S, Yamada M, Thanonkeo P. (2023) Optimization Condition for Ethanol Production from Sweet Sorghum Juice by Recombinant *Zymomonas mobilis* Overexpressing *groESL* Genes. *Energies*. 16(14):5284.
18. Mukesh Saini, Li-Jen Lin, Chung-Jen Chiang, Yun-Peng Chao. (2017). Effective production of n-butanol in Escherichia coli utilizing the glucose–glycerol mixture, Journal of the Taiwan Institute of Chemical Engineers, Volume 81, Pages 134-139, ISSN 1876-1070
19. Bosman, C. E., McClelland Pott, R. W., & Bradshaw, S. M. (2022). A Thermosiphon Photobioreactor for Photofermentative Hydrogen Production by *Rhodopseudomonas palustris*. *Bioengineering (Basel, Switzerland), 9*(8), 344.
20. Bellido-Pedraza, C. M., Torres, M. J., & Llamas, A. (2024). The Microalgae *Chlamydomonas* for Bioremediation and Bioproduct Production. *Cells, 13*(13), 1137
21. Ehsan Khorshidi Nazloo, Moslem Danesh, Mohammad-Hossein Sarrafzadeh, Navid Reza Moheimani, Houda Ennaceri. (2024). Biomass and hydrocarbon production from Botryococcus braunii: A review focusing on cultivation methods, Science of The Total Environment, Volume 926, 171734, ISSN 0048-969
22. Takeno, S., Takasaki, M., Urabayashi, A., Mimura, A., Muramatsu, T., Mitsuhashi, S., & Ikeda, M. (2013). Development of fatty acid-producing Corynebacterium glutamicum strains. *Applied and environmental microbiology, 79*(21), 6776–6783.
23. Ai-Qun Yu, Nina Kurniasih Pratomo Juwono, Jee Loon Foo, Susanna Su Jan Leong, Matthew Wook Chang, Metabolic engineering of Saccharomyces cerevisiae for the overproduction of short branched-chain fatty acids, Metabolic Engineering, Volume 34, 2016, Pages 36-43, ISSN 1096-7176

24. Baksi, S., Saha, D., Saha, S. *et al.* (2023). Pre-treatment of lignocellulosic biomass: review of various physico-chemical and biological methods influencing the extent of biomass depolymerization. *Int. J. Environ. Sci. Technol.* **20**, 13895–13922.
25. Wang, M., Ye, X., Bi, H. *et al.* (2024). Microalgae biofuels: illuminating the path to a sustainable future amidst challenges and opportunities. *Biotechnol Biofuels* **17**, 10.
26. Akshay R. Mankar, Ashish Pandey, Arindam Modak, K.K. Pant, Pretreatment of lignocellulosic biomass: A review on recent advances, Bioresource Technology, Volume 334, 2021, 125235, ISSN 0960-8524
27. Kamran Malik, Priyanka Sharma, Yulu Yang, Peng Zhang, Lihong Zhang, Xiaohong Xing, Jianwei Yue, Zhongzhong Song, Lan Nan, Su Yujun, Marwa M. El-Dalatony, El-Sayed Salama, Xiangkai Li. (2022). Lignocellulosic biomass for bioethanol: Insight into the advanced pretreatment and fermentation approaches, Industrial Crops and Products, Volume 188, Part A, 115569, ISSN 0926-6690
28. Lin Z, Cong W, Zhang J. (2023) Biobutanol Production from Acetone–Butanol–Ethanol Fermentation: Developments and Prospects. *Fermentation.* 9(9):847.
29. Choi, K. R., Jang, W. D., Yang, D., Cho, J. S., Park, D., & Lee, S. Y. (2019). Systems Metabolic Engineering Strategies: Integrating Systems and Synthetic Biology with Metabolic Engineering. *Trends in biotechnology*, 37(8), 817–837.
30. Amita Tanwar, Shashi Kumar, Vijai Singh, Pawan K. Dhar. (2020). Genome editing of algal species by CRISPR Cas9 for biofuels, Genome Engineering via CRISPR-Cas9 System, Academic Press, Pages 163-176, ISBN 9780128181409
31. Konstantinos Asimakopoulos, Hariklia N. Gavala, Ioannis V. Skiadas, Reactor systems for syngas fermentation processes: A review, Chemical Engineering Journal, Volume 348, 2018, Pages 732-744, ISSN 1385-8947
32. Carina L. Gargalo, Pau Cabaneros Lopez, Aliyeh Hasanzadeh, Isuru A. Udugama, Krist V. Gernaey, Ranjna Sirohi, Ashok Pandey, Mohammad J. Taherzadeh, Christian Larroche. (2022). On-line monitoring of process parameters during fermentation, Current Developments in Biotechnology and Bioengineering, Elsevier, Pages 117-164, ISBN 9780323911672
33. Metcalfe, G. D., Smith, T. W., & Hippler, M. (2020). On-line analysis and in situ pH monitoring of mixed acid fermentation by Escherichia coli using combined FTIR and Raman techniques. *Analytical and bioanalytical chemistry*, 412(26), 7307–7319.
34. Mizik, Tamás. (2021). Economic Aspects and Sustainability of Ethanol Production—A Systematic Literature Review. Energies. 14. 6137.
35. Hajilary, N., Rezakazemi, M. & Shirazian, S. Biofuel types and membrane separation. *Environ Chem Lett* **17**, 1–18 (2019).
36. Vasić, K., Knez, Ž., & Leitgeb, M. (2021). Bioethanol Production by Enzymatic Hydrolysis from Different Lignocellulosic Sources. *Molecules (Basel, Switzerland)*, 26(3), 753.
37. Merfort, L., Bauer, N., Humpenöder, F. *et al.* State of global land regulation inadequate to control biofuel land-use-change emissions. *Nat. Clim. Chang.* **13**, 610–612 (2023).
38. Lynd, L. The grand challenge of cellulosic biofuels. *Nat Biotechnol* **35**, 912–915 (2017).
39. Keasling, J., Garcia Martin, H., Lee, T.S. *et al.* Microbial production of advanced biofuels. *Nat Rev Microbiol* **19**, 701–715 (2021).

CHAPTER 2

MICROBIAL GROWTH DYNAMICS: UNRAVELLING THE IMPACT OF TEMPERATURE AND PH

Aditi Kumari, Ruchi Yadav, Rachna Chaturvedi, and Jyoti Prakash

Amity Institute of Biotechnology, Amity University Uttar Pradesh, Lucknow, 226028

Abstract
The importance of microbial growth and its impact on different sectors are examined in this review study. Extremophiles in particular are microorganisms that have adapted to survive in hostile environments and produce special enzymes. These enzymes are used in many different industries, such as bioremediation, medicines, and food processing. The study explores the effects of pH and temperature on pathogen virulence, enzyme stability, and microbial proliferation. It also emphasizes how temperature- and pH-tolerant microorganisms can be used in industrial settings.

Keywords:
microbial growth, enzymes, extremophiles, temperature, pH, pathogen virulence, industrial applications.

INTRODUCTION

In many biological and commercial processes, microorganisms are essential. They're useful for a lot of different things because they can generate enzymes, which are biocatalysts. The unique characteristics of microbes' enzymes, particularly those that flourish in severe environments, include stability and activity. In this review study, we will examine the significance of microbial growth, the effects of pH and temperature on microbial activity, and the possible industrial applications of microbial growth.

IMPORTANCE OF MICROBIAL GROWTH

Enzymes are biocatalysts that are essential to biochemical and metabolic processes. Since microorganisms can be cultivated in great quantities in a short amount of time and their ability to produce enzymes enhanced via genetic manipulation of bacterial cells, they are the principal source of enzymes. Furthermore, because microbial enzymes are more stable and active than those found in plants or animals, they have received greater research interest. The majority of microbes are harmful to other microorganisms because they are unable to develop and create enzymes in severe settings. Nevertheless, several microorganisms have undergone unique modifications that enable them to thrive and produce enzymes in harsh settings. Recently, a number of research initiatives have been launched to isolate novel strains of bacteria and fungi from extreme conditions, including high salinity, temperature, pH, heavy metals, and organic solvents, in order to produce various enzymes with the capacity to deliver higher [1]. The cultivation, processing, storage, and preservation of food all heavily rely on microbiology. microorganisms such bacteria, molds, and yeasts are used in the production of food and food items, including wine, beer, bakery goods, and dairy products. Yet, one of the main theories for food loss is currently believed to be the growth and contamination of hazardous and spoiling microorganisms. Food spoiling and pathogen microbe activity persist, despite the fact that technology, sanitary practices, and traceability are crucial elements in preventing and delaying microbial development and contamination[2].

TEMPERATURE EFFECT ON THE MICROBIAL GROWTH

MICROBIAL CATEGORIES BASED ON TEMPERATURE: -

- **PYSCHROPHILES:** The production of cold-active enzymes by psychrophilic/psychrotolerant microorganisms is one of their many functions. Industrial uses for psychrophile enzymes have grown in interest, in part because to continued actions taken to reduce energy usage. These icy-active enzymes offer possibilities for researching the adaptability of the possibility of biotechnological life at low temperatures abuse. The majority of research on psychrophilic bacteria has been done on cold-active enzymes such cellulase, protease, lipase, pectinase, xylanase, and amylase, chitinase, β-glucosidase, and β-galactosidase. *Acinetobacter, Aquaspirillium, Arthrobacter, Bacillus, Carnobacterium, Clostridium, Cytophaga, Flavobacterium, Marinomonas, Moraxella, Moritella, Paenibacillus, Planococcus, Pseudoalteromonas, Pseudomonas, Psychrobacter, Shewanella, Vibrio, and Xanthomonas* are among the psychrophilic microbes that produce coldactive enzymes [3].

- **MESOPHILES:** Most mesophiles typically produce easy-to-degrade enzymes under non-optimal conditions (20–40 °C, pH 6–8), with suitable amounts of substrate and growth media. The Archeabacteria are the common ancestor of certain mesophilic bacterial species that can withstand harsh environments. Modern mesophiles are thought to have separated from their archaic predecessors by extensive horizontal gene transfer and subsequent adaptation to colder climates. Mesophiles could develop and flourish at lower temperatures than extremophiles, which remained mostly genetically similar to archaic species [4].

- **THERMOPHILES AND HYPERTHERMOPHILES:** Coastal thermal springs, volcanic settings, hot springs, mud pots, fumaroles, geysers, and even deep-sea hydrothermal vents are among the places on Earth where thermophiles and hyperthermophiles can be found. In addition, they can be found in artificial settings like spray dryers, reactors, and heated composting sites. It is possible to address environmental degradation and the need for biofuels by using thermophiles and hyperthermophiles, as well as the bioproducts they produce, to support a range of industrial, agricultural, and therapeutic applications. As demonstrated by the availability of over 120 whole genome sequences for the hyperthermophiles *Aquificae, Thermotogae, Crenarchaeota, and Euryarchaeota*, efforts to sequence the entire genomes of thermophiles and hyperthermophiles are intensifying quickly [5].

- **MECHANISM OF TEMPERATURE IMPACT:** The growth and pathogenicity of a wide range of microbial species, including bacteria, fungi, viruses, and parasites, are controlled by temperature, an important and pervasive environmental signal. Microbial survival depends on their ability to respond appropriately to the cellular stress brought on by extreme temperature changes in the environment. The development and pathogenicity of microbial diseases are frequently linked to their ability to sense the physiological temperatures of their hosts. Therefore, in order to sense temperature variations, microorganisms have evolved a wide range of molecular techniques. Almost all molecules found in cells, such as proteins, lipids, RNA, and DNA, can function as thermosensors, detecting changes in the surrounding temperature and triggering pertinent biological reactions [6].

PH EFFECT ON THE MICROBIAL GROWTH

MICROBIAL CATEGORIES BASED ON PH:

- **ACIDOPHILES:** Microorganisms residing in the Archaea, Bacteria, and Eukarya domains of life are known as acidophiles; these organisms function best in environments with a pH of 3 or below. As a result of mining or natural weathering of sulfide minerals, respectively, acid mine drainage (AMD) and acid rock drainage (ARD) are two of the most acidic ecosystems on Earth. Oxygen, water, chemolithoautotrophic bacteria, and archaea all cause metal sulfides to oxidize, releasing extremely acidic effluents from the minerals that are richer in hazardous metals and metalloids. In these conditions, acidophiles have so evolved to survive not only at very low pH values but also at high metal concentrations. In order to maintain a circumneutral internal pH and numerous effective metal and metalloid resistance systems, these microbes have thereby evolved networked cellular adaptations [7].
- **ALKALOPHILES:** Though they encompass a wide range of bacterial genera and physiological kinds, alkaliphilic bacteria are ubiquitous extremophiles that face similar difficulties with cytoplasmic pH homeostasis and related issues with bioenergetic function. Alkaliphilic bacteria's biodiversity was investigated by sampling a variety of ecological niches, including nonalkaline and alkaline environments. It is noteworthy that alkaliphilic bacteria, which belong to the Gram-positive group and exhibit a variety of cultural and morphological variants, were found in all the ecosystems, including acidic soil. Several of the isolates had various enzyme activity at alkaline pH, including lipase, amylase, protease, and cellulase. The internal pH of the two potential isolates, *Bacillus lehensis* strain SB-D and *Bacillus halodurans* strain SB-W, was two units lower than the exterior pH, and they required sodium to grow [8].
- **MECHANISM OF PH IMPACT:** The effects of pH on specific enzymes, substrate speciation, and gene expression are all combined in physiological pH optimal values for microorganisms. The pH values of microorganisms that are physiologically feasible limit the consideration of possible set-points for systems such as water treatment. Below is a summary of physiological pH optimal values and an introduction to pertinent bacteria within the N-cycling network (**figure 1**) [9].

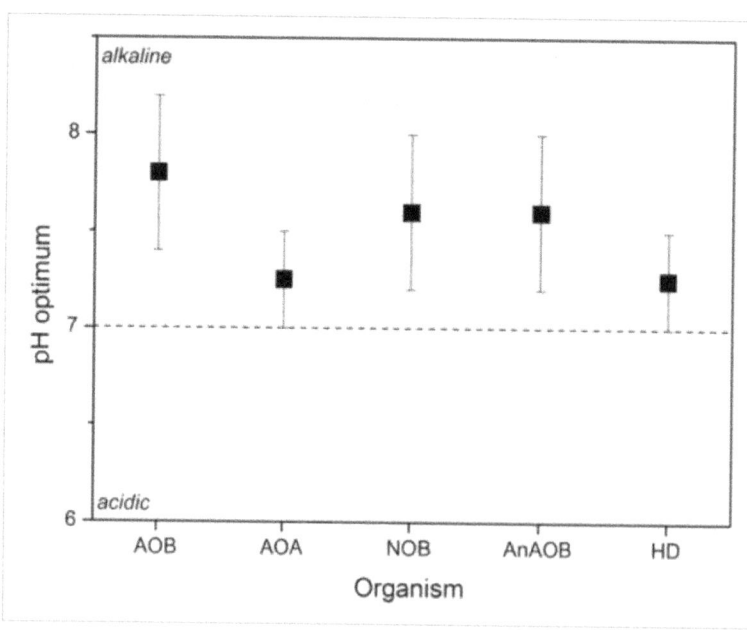

Figure 1: The ideal pH for an organism's physiological state as documented in literature. The following organisms are classified as heterotrophic denitrifiers (HD), nitrite-oxidizing bacteria (NOB), aerobic ammonia-oxidizing bacteria (AOB), (AnAOB) anaerobic ammonia oxidizing bacteria and ammonia-oxidizing archaea (AOA) [9].

ADAPTATION MECHANISMS TO TEMPERATURE AND pH EXTREMES

Comparing archaeal membranes to bacterial membranes, numerous investigations have demonstrated that the former have stronger characteristics. Yet, as widespread as mild settings are, so too do harsh environments support archaea and bacteria. A list of the minimum and maximum growth requirements for well-characterized hyperthermophiles and psychrophiles was created in order to shed light on the presence of both domains in harsh settings. The investigation revealed that both bacteria and archaea have a similarly wide temperature growth range of about 120 degrees. The dominance at the extremes of temperature, however, is a significant distinction. Higher temperature ecosystems are dominated by archaea, while lower temperature ecosystems are reported to be dominated by bacteria Another similar pattern was observed, indicating that bacteria are somewhat benefited at high alkalinity levels >pH 11, whereas archaea predominate at pH values below 1. To overcome the hostile environment, both domains significantly alter the composition of their membranes, regardless of the basic core lipid content [10].

INFLUENCE OF TEMPERATURE AND pH ON ENZYME STABILITY AND ACTIVITY

Phytases, also called myo-inositol hexakisphosphate phosphohydrolases, are a unique class of phosphatases that can hydrolyze phytate, or myo-inositol(1,2,3,4,5,6)-hexakisphosphate, in a sequential manner. Phytate is the main form of phosphorus stored in seeds and pollen. As a result, myo-inositol pentakis-, tetrakis-, tris-, bis-, and monophosphates are formed step-by-step. Orthophosphate is released during this successive hydrolysis process. Phytases are found in many different parts of nature, including plants, some animal tissues, and microbes, especially fungi. Recent years have seen a significant amount of research on phytoses due to the growing interest in using these enzymes to lower the phytate content of food intended for human consumption as well as animal feed. By releasing orthophosphate, phytotase was first suggested as an addition for animal feed to increase the nutritional value of plant material, and according to the research the following figures (2,3) have concluded the role of temperature and pH affecting phytases [11].

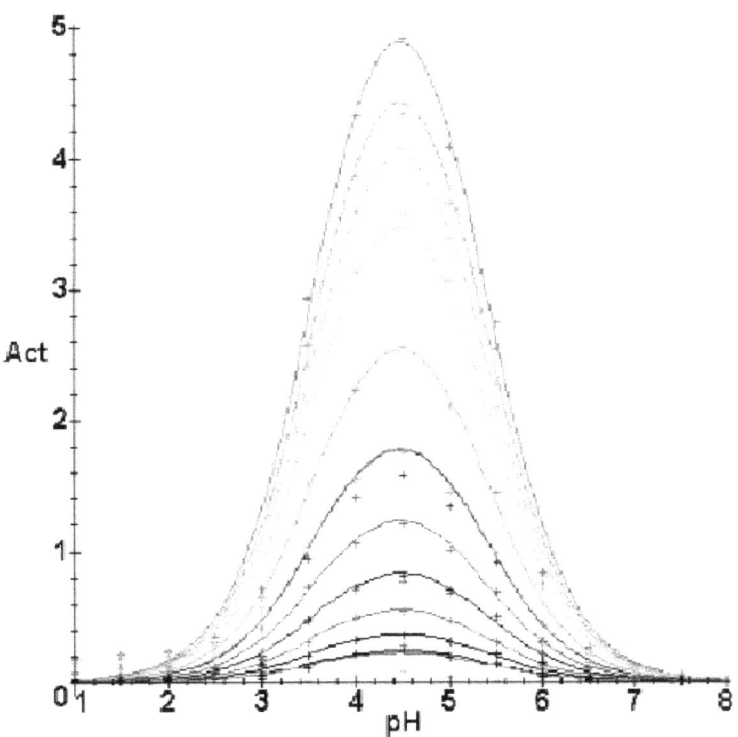

Figure 2: Activity for the phytase generated by E. coli was measured (symbols) and simulated (lines) [11].

Act = activity of enzyme 4 k_s. En (nkat)
En = concentration of active enzyme (mol/L)
k = rate constant (depending on the reaction mechanism: first order, 1/min; second order, mol/L per minute)

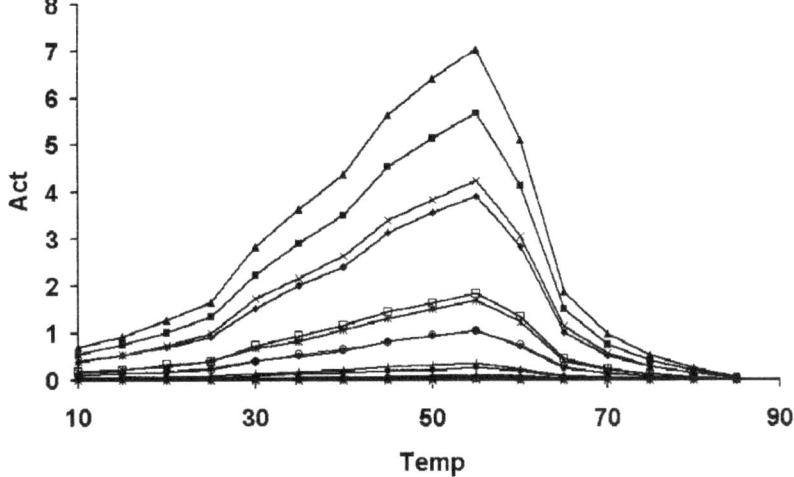

Figure 3: Phytase activity about temperature [11].

ROLE OF TEMPERATURE AND pH IN PATHOGEN VIRULENCE AND INFECTIVITY

One of the many environmental cues that bacteria may sense is temperature. Although an abrupt increase in temperature indicates danger, infections frequently use temperature variations to detect their environment and become more virulent. The ability of harmful bacteria to detect even minute temperature changes when transferring across environments is the foundation for the effectiveness of this regulation technique. Temperature-regulated virulence genes encode effectors important in attachment to host cells, motility, formation of biofilms, resistance, and immune evasion, among other major phases in the pathogenic process. A vast variety of temperature-sensing mechanisms including all the major biological macromolecules, such as protein, DNA, and RNA, have evolved during the course of evolution [12]. For fungi to survive, they must be able to adjust to changes in the pH of their host, and fungal diseases have the ability to actively alter the pH of their surroundings. Fungal proteases are expressed and activated more when the host tissues become acidified. A lot of fungi produce ammonia through nitrogen or carbon metabolism pathways, which is then expelled from the cell to increase the pH outside of the cell. The creation of an alkaline pH promotes fungal morphogenetic and reproductive activities, which are essential for the advancement of disease. These processes include hyphal growth, fruiting body formation, and germination. Because of its ability to penetrate host surfaces more easily and impede or elude immune responses, the alkaline pH promotes fungal virulence. The capacity to regulate extracellular pH is, thus, a crucial component of fungal physiology that enhances host fitness [13].

INDUSTRIAL APPLICATIONS OF TEMPERATURE AND pH TOLERANT MICROBES

Scientists are focusing on psychrotolerant bacteria because of their potential to produce biotechnological goods. The current investigation focuses on the variety of pH-tolerant and cold-tolerant isolates of Penicillium spp. and how well-suited they are to generate cold-active lipases. Using a polyphasic approach (morphological and molecular techniques), the fungal isolates were characterized. Temperature ranges of 4-35 °C (opt. 21-25 °C) and pH 2-14 (opt. 5-7) were found to be tolerated by the isolates. Temperature effects between 5 and 35 °C were studied in relation to lipase production. After 25 days of incubation, it was discovered that the fungal isolates could manufacture lipase at various temperatures. Between 15 and 25 °C was the maximum lipase production measured, and between 5 and 35 °C was the lowest. Maximum lipase production was seen at 15 °C in seven isolates (GBPI_P8, GBPI_P36, GBPI_P72, GBPI_P101, GBPI_P141, GBPI_P188, and GBPI_P222), whereas three fungal isolates (identified as GBPI_P98, GBPI_P150, and GBPI_P228) produced lipase at optimal levels at 25 °C [14].

ANTAGONISTIC EFFECT ON MICROBIAL GROWTH

Temperature and pH have an impact on the relationship between *Trichoderma sp.* and *R.solani*. The Td85 strain of Trichoderma was shown to be the most efficient strain, exhibiting the highest percentage of inhibition at 30°C, while the Td50 strain displayed reduced inhibition at same temperature. Furthermore, 4.5 was determined to be the ideal pH for the studied strains' highest antagonistic potential [15].

CONCLUSION

An important process that has broad applications in many different domains is microbial proliferation. Uniquely shaped enzymes have been found as a result of microbes' capacity to adapt to harsh environments, such as high pH or temperatures. Bioremediation, pharmaceuticals, and food processing are just a few of the businesses that use these enzymes. For industrial operations to be as efficient as possible, it is imperative to comprehend how temperature and pH affect microbial growth and enzyme activity. Find and describe new microbial strains that could be useful in developing fields should be the main goal of future study.

REFERENCES

1. Anbu, P., et al., *Microbial enzymes and their applications in industries and medicine 2016*. BioMed research international, 2017. **2017**.
2. Lorenzo, J.M., et al., *Main groups of microorganisms of relevance for food safety and stability: General aspects and overall description*, in Innovative technologies for food preservation. 2018, Elsevier. p. 53-107.
3. Yadav, A.N., et al., *Extreme cold environments: a suitable niche for selection of novel psychrotrophic microbes for biotechnological applications*. Adv Biotechnol Microbiol, 2017. **2**(2): p. 1-4.
4. Suresh, A., et al., *Recent advancements in the synthesis of novel thermostable biocatalysts and their applications in commercially important chemoenzymatic conversion processes*. Bioresource Technology, 2021. **323**: p. 124558.
5. Urbieta, M.S., et al., *Thermophiles in the genomic era: Biodiversity, science, and applications*. Biotechnology Advances, 2015. **33**(6): p. 633-647.
6. Shapiro, R.S. and L.E. Cowen, *Thermal control of microbial development and virulence: molecular mechanisms of microbial temperature sensing*. MBio, 2012. **3**(5): p. 10.1128/mbio. 00238-12.
7. Mirete, S., V. Morgante, and J.E. González-Pastor, *Acidophiles: diversity and mechanisms of adaptation to acidic environments*. Adaption of microbial life to environmental extremes: novel research results and application, 2017: p. 227-251.
8. Borkar, S., *Alkaliphilic bacteria: diversity, physiology and industrial applications*. Bioprospects of Coastal Eubacteria: Ecosystems of Goa, 2015: p. 59-83.
9. Blum, J.M., et al., *The pH dependency of N-converting enzymatic processes, pathways and microbes: effect on net N2O production*. Environmental Microbiology, 2018. **20**(5): p. 1623-1640.
10. Siliakus, M.F., J. van der Oost, and S.W. Kengen, *Adaptations of archaeal and bacterial membranes to variations in temperature, pH and pressure*. Extremophiles, 2017. **21**: p. 651-670.
11. Tijskens, L., et al., *Modeling the effect of temperature and pH on activity of enzymes: the case of phytases*. Biotechnology and bioengineering, 2001. **72**(3): p. 323-330.
12. Roncarati, D., A. Vannini, and V. Scarlato, *Temperature sensing and virulence regulation in pathogenic bacteria*. Trends in Microbiology, 2024.
13. Vylkova, S., *Environmental pH modulation by pathogenic fungi as a strategy to conquer the host*. PLoS pathogens, 2017. **13**(2): p. e1006149.

14. Pandey, N., et al., *Temperature dependent lipase production from cold and pH tolerant species of Penicillium.* Mycosphere, 2016. 7(10): p. 1533-1545.
15. PETRIȘOR, C., A. PAICA, and F. CONSTANTINESCU, *Temperature and pH influence on antagonistic potential of Trichoderma sp. strains against Rhizoctonia solani.* Sci. Papers Ser. B Hortic, 2016. **60**: p. 275-278.

CHAPTER: 3

BACTERIAL CONTRIBUTIONS TO ANTIBIOTIC DEVELOPMENT AND PRODUCTION

Alka Manisha, Ruchi Yadav, Jyoti Prakash, and Rachna Chaturvedi

Amity Institute of Biotechnology, Amity University Uttar Pradesh, Lucknow Campus, Lucknow- 227105, Uttar Pradesh, India.

Abstract

It is recognised that many bacteria, or microorganisms, create a wide range of antibiotics that are manufactured and utilised to treat sickness and potentially fatal conditions. Numerous tiny organisms, including bacteria, fungus, and actinomycetes, manufacture antibiotics as part of their defence mechanism against nearby microorganisms. Bacteria have evolved intricate biochemical pathways to synthesize a diverse array of secondary metabolites, including antibiotics. These compounds serve as chemical warfare agents, providing bacteria with a competitive advantage in their natural environments. Over millennia, humans have harnessed this natural ability, cultivating and optimizing bacterial strains to produce life-saving antibiotics on an industrial scale. Technology and conventional programs have been used to create a range of kinds based on random mutation and testing, improving the yield of antimicrobials in the industry. Increased production and the biosynthetic processes that produce novel antibiotics have been made possible by the invention of DNA fusion methods and their use in microorganisms that produce antibiotics. This review article will concentrate on the role that various bacteria play in the production of many antibiotics.

Keywords:
Antibiotics, Metabolites, Antimicrobial, Biochemical.

INTRODUCTION

Antibiotics are a class of antimicrobial agents designed to combat bacterial infections. The term "antibiotic" originates from the Greek words *anti-* meaning "against" and *bios* meaning "life," which together signify "against life." This is because antibiotics specifically target living bacterial cells. Antibiotics are typically small, light-weight organic molecules, produced either naturally by microorganisms like fungi and bacteria or synthetically through chemical processes. In nature, microorganisms produce antibiotics as a defense mechanism against other competing microbes in their environment. [1].Antibiotics work by targeting specific structures or processes within bacterial cells, interfering with their growth, reproduction, or survival [2]. Some antibiotics prevent bacteria from forming their cell walls, while others impair vital metabolic functions or prevent the creation of proteins required for bacterial life. Antibiotics are effective therapies for bacterial infections because they specifically are a useful resource for screening and finding novel antibiotics. Microbiologists investigate a variety of habitats, including soil, water, and even the human microbiome, to extract bacteria and investigate their antimicrobial capabilities. [3].In order to battle drug-resistant bacteria and tackle the ongoing problem of antimicrobial resistance, scientists are using the genetic diversity of bacteria to develop novel chemicals. Antibiotic production and secretion by bacteria has evolved as a defence strategy against competing pathogens or microorganisms. [4] Not only can knowing how bacteria produce antibiotics benefit in the creation of novel antibiotics, but it also helps fight antibiotic resistance. Bacteria play many roles in the research from antibiotic production to gene expression and metabolic pathway. Bacteria play a crucial role in the production of antibiotics. Many antibiotics are derived from natural compounds produced by bacteria themselves. These compounds help bacteria survive by inhibiting or killing competing microorganisms in their environment. Here's a breakdown of how bacteria contribute to antibiotic production:

Natural Producers of Antibiotics
- **ACTINOMYCETES:** This group of bacteria, especially *Streptomyces* species, is the source of many antibiotics, including streptomycin, tetracycline, and erythromycin. *Streptomyces* bacteria produce these antibiotics as secondary metabolites, which are compounds not directly involved in the normal growth, development, or reproduction of the organism.
- **BACILLUS SPECIES:** These bacteria produce antibiotics such as bacitracin and polymyxin. Bacillus species are found in soil and produce these antibiotics to outcompete other microbes in their environment.

Biotechnological Production
- **FERMENTATION:** In industrial settings, bacteria are grown in large fermentation tanks under controlled conditions to produce antibiotics on a large scale. During fermentation, bacteria convert nutrients into antibiotics, which are then extracted and purified.
- **GENETIC ENGINEERING:** Modern biotechnology allows for the modification of bacterial strains to enhance antibiotic production or to produce novel antibiotics. This is done by introducing or altering genes responsible for antibiotic biosynthesis.

Antibiotic Resistance Research
- **DISCOVERY OF RESISTANCE GENES:** Bacteria are also used to study antibiotic resistance, which can lead to the development of new antibiotics. By understanding how bacteria develop resistance, scientists can design antibiotics that are less susceptible to resistance mechanisms.
- **Screening for New Antibiotics:** Soil bacteria are often screened for new antibiotic-producing capabilities. These bacteria can produce new compounds that might be effective against resistant strains of bacteria.

Bacterial Contributions to Antibiotic Development and Production

Symbiotic Relationships

Some bacteria produce antibiotics as part of a symbiotic relationship with plants or animals, protecting their host from pathogenic microbes. For example, certain bacteria living in the soil around plant roots produce antibiotics that protect the plant from fungal infections.

Antibiotic Producing Bacteria

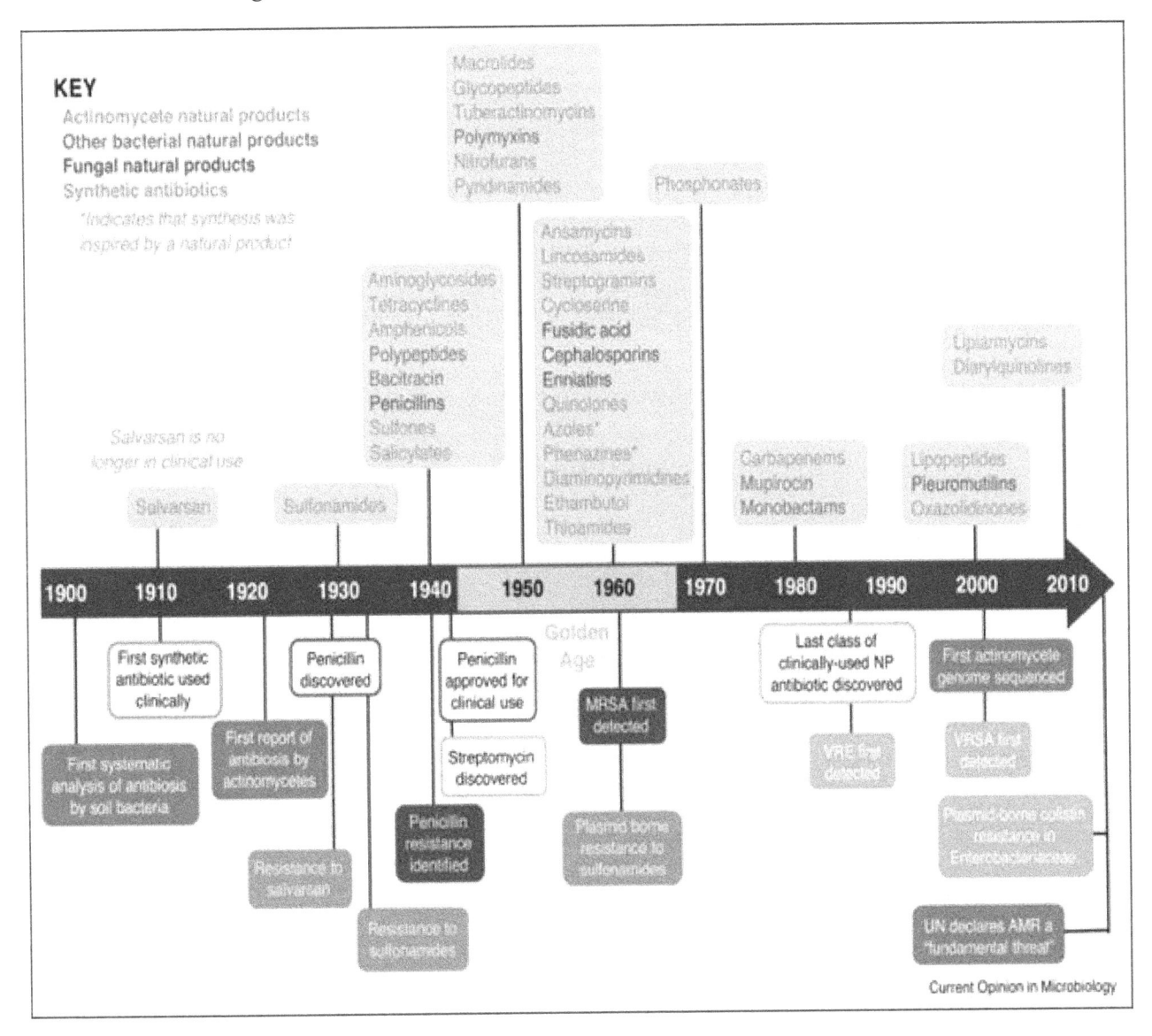

Figure 1: Some natural products of bacteria are mentioned above.

Different antibiotics are available and are used to treat various infections. When a patient is in the hospital with a serious infection, antibiotics can be administered intravenously or taken by mouth as a liquid, capsule, or tablet. [4]

Some common antibiotics produced by bacteria are: -
1. Penicillin – for example Penicillin v., Fluclorocillin and Amoxicillin.
2. Tetracycines – for example Tetracycline, Doxycycline and Minocycline.
3. Cephalosporins – for example Cefaclor, Cefadroxil and Cephalexin.
4. Macrolides – for example Erythromycin, Azithromycin, and Clarithromycin.
5. Aminoglycosides – for example Gentamycin, Amikacin and Tobramycin.

Table 1: Antibiotic produced by bacteria0

S.No	Antibiotics	Bacterial strains producing antibiotic
1.	Streptomycin	Streptomyces griseus
2.	Gentamicin	Micromonospora purpurea
3.	Polymyxin B	Bacillus polymyxa
4.	Bacitracin	Bacillus licheniformis
5.	Erythromycin	Saccharopolyspora erythraea.

ANTIBIOTICS PRODUCED BY ACTINOMYCETES

According to data from 16s ribosomal cataloguing and DNA rRNA pairing investigations, actinomycetes are gram-positive bacteria with a high Guanine (G) to Cytosine (C) ratio in their DNA (55 mol%) that are phylogenetically related. [5]

A group of bacteria known as the Actinomycetes has numerous significant and unique characteristics. As manufacturers of antibiotics and other therapeutically beneficial chemicals, they are extremely valuable. They display a variety of life cycles that are distinct from those of other prokaryotes, and they seem to be crucial to the cycling of organic materials in the soil ecosystem. [1] Actinomycetes have been more well-known in recent years due to their ability to produce antibiotics. [6].Antibiotics produced by actinomycetes are Streptomycin, Gentamycin, Bacitracin, B-Polymoxin, erythromysin some of the antibiotics which are in use presently and is the product of Actinomycetes.

STREPTOMYCIN

A variety of settings support the growth of the Gram-positive bacterium streptomyces, which has a filamentous fungus-like structure. Streptomyces has morphological variations that involve the development of a hyphae layer that can separate into a string of letters. A Gram-positive bacteria with the same filamentous shape as mould, streptomyces, develops in many environments. [7] With the discovery of streptothricin in 1942, the history of antibiotics derived from Streptomyces began. Since the beginning of our research on actinomycetes, Actinomyces, a kind of bacteria, have demonstrated how to produce and distinguish themselves from the species of this other group of chemical compounds that contain antibodies. Streptothricin was a novel chemical that was quickly phased out in 1942. It was non-toxic to animals and effective against both gram positive and gram negative bacteria. [8]

Figure 2: Antibiotic produced by Streptomyces [5]

- Platensimycin 2006 *S. platensis*
- Daptomycin 2003 *S. roseosporus*
- Linezolid 2000 Synthetic
- Mupirocin 1985 *Pseudomonas fluorescens*
- Ribostamycin 1970 *S. ribosidificus*
- Fosfomycin 1969 *S. fradiae*
- Trimethoprim 1968 Synthetic
- Gentamicin 1963 *Micromonospora purpurea*
- Fusidic acid 1963 *Fusidium coccineum*
- Nalidixic acid 1962 Synthetic
- Tinidazole 1959 Synthetic
- Kanamycin 1957 *S. kanamyceticus*
- Rifamycin 1957 *Amycolatopsis mediterranei*
- Noviobiocin 1956 *S. niveus*
- Vancomycin 1956 *S. orientalis*
- Cycloserine 1955 *S. garyphalus*
- Lincomycin 1952 *S. lincolnensis*
- Erythromycin 1952 *Saccharopolyspora erythraea*
- Virginiamycin 1952 in *S. pristinaespiralis S. virginiae*
- Isoniazid 1951 Synthetic
- Viomycin 1951 *S. vinaceus e S. capreolus*
- Isoniazid 1951 Synthetic
- Viomycin 1951 *S. vinaceus e S. capreolus*
- Nystatin 1950 *S. noursei*
- Tetracycline 1950 *S. aureofaciens*
- Neomycin 1949 *S. fradiae*
- Chloramphenicol 1949 *S. venezuelae*
- Polymyxin 1947 *Bacillus polymyxa*
- Nitrofurantoin 1947 Synthetic
- Cephalosporins 1945 *S. clavuligerus*
- Bacitracin 1945 *Bacillus licheniformis*
- Cephalosporins 1945 *S. clavuligerus*
- Streptomycin 1944 *S. griseus*
- Penicillin 1941 *Penicillium chrysogenum*

Belonging to the aminoglycoside family of antibiotics, streptomycin was the first to be demonstrated to be effective against tuberculosis (TB). Broad-spectrum action, especially against gram-negative bacteria, is demonstrated by the naturally occurring bacterium, Streptomyces griseus.

Mechanism of Action:

Streptomycin inhibits protein synthesis by attaching itself to the bacterial ribosome's 30S subunit. Protein synthesis mistakes are brought about by this binding's interference with the translation process. These errors result in the synthesis of harmful or nonfunctional proteins, which ultimately cause the death of bacterial cells.

Streptomycin has the same bactericidal effect as other aminoglycosides, which means that it kills bacteria directly as opposed to only preventing their development. [5].Mycobacterium tuberculosis: The first antibiotic discovered to be effective against the TB-causing Mycobacterium tuberculosis.

In the treatment of tuberculosis, it is still a crucial medication, especially for strains that are resistant to conventional first-line treatments.

Additionally, it has activity against a variety of gram-negative bacteria, such as Escherichia coli and Klebsiella.

GENTAMYCIN

Gentamicin is a broad-spectrum antibiotic that belongs to the aminoglycoside class. It is primarily used to treat severe bacterial infections caused by gram-negative bacteria and certain gram-positive bacteria. Soil-dwelling bacteria belonging to the genus Micromonospora are responsible for producing gentamicin. Micromonospora purpurea is the species that produces gentamicin the most commonly. In order to protect themselves against the proliferation of rival microorganisms in their surroundings, these bacteria naturally produce gentamicin. [9]

Mechanism of Action
Protein Synthesis Inhibition: Gentamicin works by binding to the bacterial 30S ribosomal subunit, which is part of the machinery responsible for protein synthesis. This binding interferes with the translation process, leading to the production of faulty proteins and ultimately causing bacterial cell death.

Spectrum of Activity
- **Gram-Negative Bacteria:** Gentamicin is particularly effective against gram-negative bacteria such as *Escherichia coli, Klebsiella pneumoniae, Pseudomonas aeruginosa,* and *Serratia marcescens*.
- **Gram-Positive Bacteria:** It is also used to treat infections caused by some gram-positive bacteria, such as *Staphylococcus aureus*, including methicillin-resistant *Staphylococcus aureus* (MRSA), but typically in combination with other antibiotics. [10]

Biosynthesis Pathway
- **Complex Biosynthetic Process:** The production of gentamicin by *Micromonospora* species is a complex process involving multiple enzymatic steps. The biosynthesis begins with the formation of a core aminocyclitol ring structure, which is then glycosylated with various sugar moieties to form the active gentamicin molecule.
- **Gene Clusters:** The genes responsible for gentamicin biosynthesis are typically organized in clusters within the bacterial genome. These gene clusters encode the enzymes needed for each step of the antibiotic's synthesis, from precursor formation to the final assembly of the gentamicin molecule.

ERYTHROMYCIN

A polyketide antibiotic called erythromycin is made mostly in submerged culture by Saccharopolyspora erythraea. This antibiotic has a wide range of medical uses in the treatment of numerous illnesses brought on by gram-positive and some gram-negative bacteria. Erythromycin is therefore frequently used in clinical settings to treat respiratory infectious illnesses in the treatment of many acute infections caused by bacteria belonging to *Staphylococci* spp. and *Neisseria* spp and as an antimalarial in combination with other drugs to reduce the pathogen resistance. This antibiotic is mostly produced using a submerged culture technique containing either free cells of cultured bacteria. [7]

Role of Bacteria in Erythromycin Production

- **Biosynthesis route:** Saccharopolyspora erythraea uses a complex metabolic route to produce erythromycin. The bacterium's genome contains a cluster of important biosynthetic genes that assemble erythromycin step-by-step from the less complex precursor. [11]
- **Polyketide Synthase (PKS):** A multi-enzyme complex known as polyketide synthase (PKS) is involved in the production of erythromycin. Through a sequence of condensation, reduction, and cyclization processes, the PKS machinery assembles erythromycin from tiny molecules such as propionyl-CoA and methylmalonyl-CoA. [5]
- **Production of Secondary Metabolite:** Because it is created during the stationary phase of bacterial development, when resources become limited and the bacterium switches its metabolic focus from growth to producing bioactive molecules, erythromycin is categorized as a secondary metabolite. [12]

BACITRACIN

Bacitracin is an antibiotic commonly used in topical ointments to prevent infections in minor cuts, burns, and scrapes. It works by inhibiting the synthesis of bacterial cell walls, making it effective against a variety of gram-positive bacteria. Bacillus licheniformis is the most prominent naturally occurring bacitracin producer. [4]

During the growth phase, Bacillus licheniformis produces bacitracin as a secondary metabolite. The bacterium obtains a competitive advantage by the synthesis of bacitracin, which aids in inhibiting the growth of competitor bacteria in its surroundings.Bacillus subtilis: Bacillus subtilis is a different strain that can also produce bacitracin. However, because of its greater antibiotic output, Bacillus licheniformis is employed extensively in industrial settings. [9]

Biosynthesis Pathway

Bacitracin is a cyclic peptide antibiotic composed of a mixture of related compounds (bacitracin A, B, C, etc.). The biosynthesis of bacitracin in Bacillus species involves a non-ribosomal peptide synthetase (NRPS) pathway. This pathway assembles the peptide backbone of bacitracin from amino acid precursors, which are then cyclized to form the active antibiotic compound. [13]

The genes responsible for bacitracin biosynthesis are located in a specific gene cluster in the bacterial genome. These genes encode the enzymes required for the step-by-step assembly and modification of the bacitracin molecule.

Industrial Production

- **FERMENTATION PROCESS:** In an industrial setting, *Bacillus licheniformis* is cultivated in large fermentation tanks under controlled conditions to maximize bacitracin production. The bacteria are provided with a nutrient-rich medium that supports their growth and the production of bacitracin.
- **EXTRACTION AND PURIFICATION:** After the fermentation process, bacitracin is extracted from the culture medium and purified. The purified antibiotic is then formulated into various pharmaceutical products, such as ointments and creams.

B-POLYMIXIN

Polymyxin B is an antibiotic that belongs to the polymyxin class, which is primarily used to treat infections caused by gram-negative bacteria. Polymyxin B is particularly effective against bacteria that are resistant to other antibiotics, making it a valuable option in treating multidrug-resistant infections. [12] The cationic lipopeptide antibiotic family, which includes polymyxins, is produced by Gram-positive bacteria and has broad antimicrobial activity. Despite the fact that polymyxins are extremely efficient antimicrobials, their clinical application is constrained by their nature, toxicity, and the accessibility of alternative powerful but non-toxic antibiotics. Lipopolysaccharide is present in high concentrations in polymyxins, which are now part of hemoperfusion cartridges used to remove it from blood and stop the onset of sepsis. [1]

Mechanism of Action:

Polymyxin B disrupts cell membranes by attaching itself to phospholipids and lipopolysaccharides (LPS) found in gram-negative bacteria's outer membrane. The bacterial cell membrane's integrity gets damaged by this binding, making it more permeable. Consequently, the bacterial cell releases its contents, causing cell death. Polymyxin B has a bactericidal effect, which means that it kills germs directly as opposed to merely preventing their development.

Spectrum Activity

Gram-Negative Bacteria: Pseudomonas aeruginosa, Escherichia coli, Klebsiella pneumoniae, Enterobacter species, and Acinetobacter baumannii are among the gram-negative bacteria that Polymyxin B is particularly effective against. It works very well against bacteria that are resistant to several drugs.

Methods for Production of Antibiotics

Despite the large number of different antibiotics that are known, less than 1% of antimicrobial substances are useful for medicine or industry. Penicillin, for instance, has a high therapeutic index because it typically does not affect human cells, but this is not true of many antibiotics. Other antibiotics simply have no advantages over those currently in use or don't have any other useful uses.

A screening procedure is frequently used to find effective antibiotics. In order to carry out such a screen, isolates of numerous different microorganisms are cultured, and their ability to produce diffusible substances that hinder the growth of test organisms is then examined. The majority of antibiotics found in such a screen must be ignored because they are already well known. The remaining substances must be examined and possibly modified for their potential therapeutic effects and selective toxicities. [5]. A rational design programme is a more contemporary variation of this strategy. Instead of using tests to demonstrate general inhibition of a culture, this involves screening aimed at discovering new natural products that inhibit a specific target, such as an enzyme only present in the target pathogen. [14]

CONCLUSION

In today's world the drug resistant is a major threat to people's health. From hundreds to thousands of lives are lost every year only because of infection that can no longer be treated with existing drugs. To overcome the problem of drug resistant and antibiotic resistant researchers and scientists are doing research on the natural antibiotic producing bacteria which produce antibiotic as a defence against competing pathogens. They can both be genetically modified to produce particular chemicals and are the natural producers of antibiotics. Bacteria act as living factories, converting nutrients into antibiotics through fermentation processes. Bacterial bioprospecting activities aid in the discovery of novel antibiotic compounds. Additionally, within microbial communities, bacteria engage in synergistic relationships that result in the production of antibiotics that defend against pathogens. Bacteria, once solely viewed as pathogens, have emerged as invaluable allies in the fight against infectious diseases. Their capacity to produce a diverse array of antibiotics has revolutionized healthcare. This paper explored the intricate mechanisms underlying bacterial antibiotic biosynthesis, highlighting the pivotal role of key genera like Streptomyces. While antibiotics have undoubtedly saved countless lives, the escalating challenge of antibiotic resistance poses a significant threat to global health. It is imperative to intensify research efforts to discover novel antibiotics and develop strategies to combat resistance. By delving deeper into the genetic and biochemical intricacies of antibiotic production, scientists can unlock the full potential of these microbial factories.

REFERENCES

1. J. Anshu Gupta, "Isolation, Identification And Characterisation Of Antibiotic Producing Bacteria From Soil At Dr. C V Raman University Campus Bilaspur," *World Journal of Pharmaceutical Research*, 2017.
2. H. Ogawara, "Comparison of Antibiotic Resistance Mechanisms in Antibiotic-Producing and Pathogenic Bacteria," *Molecules 2019*, p. 3430, 2019.
3. Britanica, "Statin production," Britanica information, 2023.
4. D. N. S. S. K. J. G. S. V. A. Gupta K, "Plant growth promoting rhizobacteria (PGPR): current and future prospects in crop improvement.," *Current trends in microbial biotechnology for sustainable agriculture.*, pp. 203-206, 2021.
5. X. E. M. C. P. e. a. Jiang, "Dissemination of antibiotic resistance genes from antibiotic producers to pathogens," *Nat Commun*, p. 15784, 2017.
6. A. C. M. A. Amita Gaurav Dimri1, "Antibiotic potential of actinomycetes from different environments against human pathogens and microorganisms of industrial importance," *Science Archives*, pp. 7-24, 2020.
7. M.N. Pham, "Recent advancement of eliminating antibiotic resistance bacteria and antibiotic resistance genes in livestock waste: A review," *Elsevier*, vol. 26, p. 103751, 2024.
8. S. Katara, "Role of bacteria and fungi in antibiotic production," *The pharma Innovation*, pp. 709-714, 2020.
9. Mahreen Ul Hassan 1 2, "Characterisation of Bacteriocins Produced by Lactobacillus spp. Isolated from the Traditional Pakistani Yoghurt and Their Antimicrobial Activity against Common Foodborne Pathogens," *national library of medicines*, 2020.
10. C.BJ, «gentamicin,» *Europe PMC*, 2023.

11. M. Zhang, "Effects of erythromycin and sulfamethoxazole on Microcystis aeruginosa: Cytotoxic endpoints, production and release of microcystin-LR," *elsevier,* vol. 399, p. 123021, 2020.
12. S. Aghamohammad, "Antibiotic resistance and the alternatives to conventional antibiotics: The role of probiotics and microbiota in combating antimicrobial resistance," *Elsevier,* vol. 267, p. 127275, 2023.
13. C. K. Sharma and M. Sharma, "Up Scaling Strategies to Improve the Industrial Production of Bacitracin at Largescale," *Mini Reviews in Medicinal Chemistry,* pp. [1548 - 1556], 2017.
14. H.U.Heul, "Regulation of antibiotic production in Actinobacteria: new perspectives from the post-genomic era," *Natural Product Reports,* 2018.

CHAPTER: **4**

LICHEN AS BIOINDICATORS FOR AIR POLLUTION DETECTION

Amar Prakash Shukla and Jyoti Prakash

Amity Institute of Biotechnology, Amity University Uttar Pradesh, Lucknow, 226028

Abstract

Lichens are widely used as biomonitors of air pollution because these organisms respond to toxic gases released by the plant—particularly sulfur dioxide and nitrogen dioxide, at the cellular organisms, and colony levels. However, due to the botanical and environmental changes of organism biologists often have trouble seeing the effects of air pollution on background noise. On the other hand, the response is needed if we want to incorporate these ideas into environmental management decisions. In this section, we explore this complex business by focusing on the biomarker process based on the analysis of the reduction of lichen communities associated with an increase in air pollution wind. We specifically report examples of the use of bioexamination in heterogeneous environments to give an overview of important specimens, analysis, and statistical methods that can be used by researchers and stakeholders to better understand the relationship between lichen diversity and air pollution. We identified three main factors of change in lichen communities at the physical and spatial level. First, we investigated the differential responses of lichens to air pollutants associated with extreme weather conditions. Due to the close proximity of their metabolic processes to the atmosphere, lichens are strongly affected by climatic changes such as precipitation and temperature. Today's environmental assessments involve evaluating the impact of pollutants on human health and natural resources. Therefore, determination the effects of gaseous pollutants on natural ecosystems by the of lichen biomonitoring.In the next example, we represent the design of the lichen biomonitoring in the natural forest. Some geographically relevant variables will play an important role in lichen diversity and that more care should be taken when interpreting these data in terms of the direct effects of pollution. At the detailed scale, the variable lichen diversity can be too high, thus, the level of uncertainty in the interpretation of the data increases this should be taken into account when planning larger studies by allowing the use of appropriate samples where indicated. Next example, we addressing issue, showing data from extensive experiments and comparison changes in the non-weather related natural environment in the climate zone for the larger difference observed in the anthropogenic area when compared.

Keywords:
Lichens, Climate, Pollution, Biomonitors

INTRODUCTION

Lichens are thought to be the result of a combination of fungi and algae. More precisely, the word "algae" refers to the family Cyanobacteria or Chlorophyceaethe fungus is usually an Ascomycete, but can sometimes be a Basidiomycete or Phycobacterium. In this organization, the algae are partially responsible for the production of nutrients as they contain chlorophyll (Chl), while the fungus provides water and minerals to the algae. (1) This disease takes a perennial, time-like form. They are slow growing, dependent on the environment for nutrients, and - unlike vascular plants - they do not die when grown. In addition, it has no cuticles or pores, which means that different pollutants are absorbed throughout the entire organism. As early as 1866, research was published on epiphytic lichens used as biomarkers. Lichens are the most studied creatures on air quality. They are described as "permanent control" for climate measurement. (2) Over the last 30 years, many studies have highlighted the possibility of using lichens as biomonitors of air quality, because lichens are sensitive to many aspects of the environment, which may lead to changes and or certain restrictions in some of their contents. In fact, many physical parameters have been used to measure environmental damage to lichens, for example: photosynthesis; chlorophyll content and degradation; ATP decreases; changes in breathing; changes in endogenous auxin levels; and ethylene production. In addition, exposure to SO_2 causes membrane damage in lichen cells. Many studies have shown that the sulfur content in lichens has a positive relationship with the sulfur dioxide in the atmosphere. Many author have reported that Chla+b concentration vary with traffic pollution and urban emissions. Generally, lichens transferred to traffic areas showed an increase in Chl a + b proportional to emissions. (2)These effects are due to traffic emissions, particularly sulfur and nitrogen oxides. Chlorophyll b / chlorophyll a ratios are higher in areas with heavy traffic and industrial pollution. Air traffic, particularly gasoline and benzene, seemed to have a smaller effect on lichen populations than traffic. A case study at Hamburg Airport proves this. Lichens can be used in two different ways as biological and/or chemical indicators.By drawing all the animals in the field (Method A); ii. By sampling lichen species one by one and measuring the pollution in their thalli; or by transplanting lichens from uncontaminated areas to polluted areas, then measure the morphological changes of lichens. Assess thallus and/or physical weakness and/or assess bioaccumulation of contaminants (Method B).

LICHENS

Lichens are symbionts of fungi and different organisms such as cyanobacteria or algae. Cyanobacteria are also known as blue-green algae as well as different types of algae. Non-fungal products are called chlorophyll-containing photo-organisms. Many partners of lichens, including a photobiont and a fungus, are not ubiquitous, and there are sometimes lichens with more than one photobiont partner. (1)Fungal chaperones are produced by filamentous cells, each of which is called hyphae. Most fungi in lichens are ascomycetes or basidiomycetes. Green algae Blue-green organisms of the family Chlorophyta or Cyanophyceae are the most common algae. Fungal partners cannot survive without algae, which can survive alone in water or moisture. (3) The color of lichens varies from yellow to green to black.Most lichens grow slowly. Algae are blue-green organisms that convert nitrogen into ammonia. Lichens that live in stressful environments such as the arctic tundra or alpine mountains are said to live for hundreds of years.

CLASSIFICATION OF LICHENS

1. CRUSTOSE LICHENS
2. FOLIOSE LICHENS
3. FRUSTICOSE LICHENS

1. **CRUSTOSE LICHENS:-** Crustose lichens are hardy. They form concrete on surfaces such as concrete, dirt or tiles. They can come in many different colors such as sun yellow, orange and red, as well as gray and green. Crustose lichens are pressed against their substrate. (2)
2. **FOLIOSE LICHENS** Foliose patterns are made from leaf-like patterns. It can be easily removed from the place where it grows. Foliose is named for leaves that resemble plant leaves. (2)
3. **FRUTICOSE LICHENS** Fruticose is like a small tree, like a small tree without leaves. It looks like dense coral. It grows on rocks, trees and soil. The word fruticose is of Latin origin and means tree or bush. (2)

The non-fungal portion, which has chlorophyll, is referred to as a photobiont, while the fungal portion is known as a mycobiont.There are lichens that have more than one photobiont partner, and many lichen partners also comprise one universal mycobiont and one photobiont.In addition, lichens generate pigments and other secondary chemicals that have dye-like properties. (4).They are utilized in traditional medicine and as delicacy in several regions. Lead and copper contaminants as well as polyester resins can be broken down by some lichens.

Lichens typically thrive in a variety of environmental settings. However, they are highly susceptible to air contaminants such as nitrogen and sulfur. (5).Lichens can serve as indicator species since air pollution also causes a decrease in lichen populations. The non-fungal portion, which has chlorophyll, is referred to as a photobiont, while the fungal portion is known as a mycobiont.There are lichens that have more than one photobiont partner, and many lichen partners also comprise one universal mycobiont and one photobiont. (4). In addition, lichens generate pigments and other secondary chemicals that have dye-like properties.They are utilized in traditional medicine and as delicacy in several regions. Lead and copper contaminants as well as polyester resins can be broken down by some lichens. Lichens typically thrive in a variety of environmental settings. However, they are highly susceptible to air contaminants such as nitrogen and sulfur. (5). Lichens can serve as indicator species since air pollution also causes a decrease in lichen populations.

CRUSTOSE LICHEN   FRUTICOSE LICHEN   FOLIOSE LICHEN

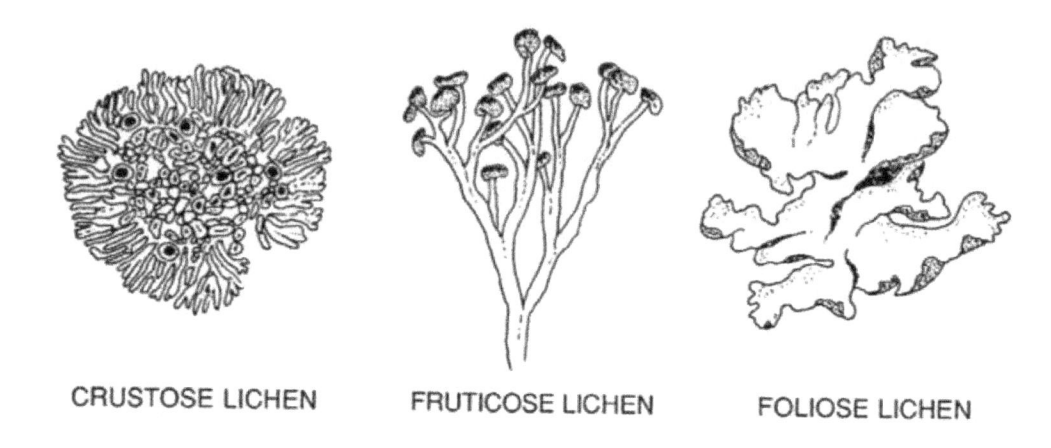

Figure 1: Types of Lichens

LICHENS ARE CALLED AS POLLUTION INDICATOR

The non-fungal portion, which has chlorophyll, is referred to as a photobiont, while the fungal portion is known as a mycobiont. (6)

Some lichens have more than one photobiont partner, and many lichen partners also comprise one universal mycobiont and one photobiont.

In addition, lichens generate pigments and other secondary chemicals that have dye-like properties. (7)

They are utilized in traditional medicine and as delicacy in several regions. Lead and copper contaminants as well as polyester resins can be broken down by some lichens.

Lichens typically thrive in a variety of environmental settings. However, they are highly susceptible to air contaminants such as nitrogen and sulfur. (8)

Lichens can serve as indicator species since air pollution also causes a decrease in lichen populations. (9)

MAJOR LICHEN-SENSITIVE AIR POLLUTANTS

There is huge variety of elements and chemical compounds present in the atmosphere that affect the growth of lichen and distribution of lichen. Pollutants are such as sulfur di oxide, nitrogen di oxide and fluoride compound (10)

- **SULFUR DIOXIDE:** sulfur dioxide dissolves in h20 to give acidic ions which are easily absorbed through the lichen thalli dismembering photosynthesis. Sulfur dioxide has also been displayed to inhibit the exertion of nitrogenase, which is used by cyanbionts to fix atmospheric nitrogen. (11)
- **NITROGEN DIOXIDE:** Lichens need nitrogen to survive, as it preserves their vital proteins and organic acids. Additionally, cyanobacteria do what's called nitrogen fixation, converting the element into a stable and simple form, ammonia. Nitrogen dioxide and nitric oxide are the nitrogen used by the lichen in the form of gas. few lichen are sensitive to ammonia and some lichens are not sensitive. In fact, more nitrogen can damage many lichen species and cause them to die. Excess nitrogen also kills the algae's chlorophyll, depriving the fungi of the sugar they need to survive. (11).

LICHENS ARE USED AS AIR QUALITY INDICATOR

In the US, the fitness of different lichen species is examined and join with climate data to show significant and overall pollution. Atmospheric precipitation is the transfer of gases and substances from the atmosphere to the earth. In this case, biologist is investigating the level of nitrogen accumulation to protect sensitive ecosystems. (12).Lichens are also used as pollution indicators in Germany because they are sensitive to many pollutants.In particular, lichens are most sensitive to sulfur dioxide, but not to other acidic compounds, including heavy metals and photo oxidants. In Germany there are two ways to use lichens as an indicator: to expose lichens and to record lichen growth on trees under certain conditions. Trees are registered according to certain rules:

CIRCUMFERENCE OF AT LEAST 60 CM, NO OR VERY LITTLE SHADE

It should be noted that many lichen species are sensitive to pollution. Therefore, the diversity of lichen species is taken into account when calculating and representing the AQI. This is happen by determining the coverage area and capacity. (13).By the Dutch developed the lichens are classifying, In high nitrogen environment and high ph the lichens are developed, while opposite in the acid plant lichen species. Shell pH has been shown to be affected by sulfur dioxide concentration. (11)This approach has been used to map and examined patterns of nitrogen and NH3 pollution across ASIA and has also helped expose the responses of lichens to global warming. The Dutch method should not be confused with the Ellenberger method, which uses the "Ellenberger N value" to estimate where the productivity/macronutrient gradient per species reaches its maximum. The system includes the "Ellenberger N-Index", which is determined by giving each plant species an N-score as the mean score of the community on a scale of 1 to 10 from poor food to food rich. (14).This approach is also suitable for showing environmental effects, as it is used to show changes and colonization over time.Lichens have also been used to monitor nitrogen levels in tea plantations and Himalayan forests in Sri Lanka. Farmers use fertilizers that contain nitrogen in the form of ammonia, so research is ongoing to understand how this season will affect land area.

ALL LICHENS INDICATE THAT THE AIR IS CLEAN

No, but most lichens are sensitive to air pollution, so they will not grow if there is high pollution. However, some lichens tolerate pollution and some can still benefit from neutral nitrogen pollution. (10)

LICHEN-BASED NITROGEN AIR QUALITY INDEX GUIDE

The FSC Lichen-Based Nitrogen Air Quality Index Guide contains direct keys to oak and birch lichens that you can use to measure local air quality and air quality changes in recent years.To measure air quality of your location, lichen index score (LIS), the score is converted into the nitrogen air quality index (NAQI) TO show air quality. (2).Lichens are the complex organism that live together and consist of bacteria and one or more species of algae. Algae actively produce essential nutrients through photosynthesis.

Fungal partners, meanwhile, provide the area where algae live.

Lichens are very quite sensitive to small changes in the environment. Therefore, they work as an indicator of the health of our environment. In the past, Industry was major pollutant by the SO_2 from burning of coal this is the main reason of acid rain. Today however, nitrogen compounds from intensive agriculture and motor vehicles have become major pollutants. Although pollution can kill many lichens, others tolerate and even live in them. Which lichens are acid and nitrogenous and which lichens grow in non-toxic areas are indicated by the GUIDE. Ancient research on English oak and birch trees has shown that lichens are sensitive to increased levels of pollutants in the air.In the recent times; Air pollution has come a major problem not only in industrialized countries, but in present time it raised as a serious environmental matter substantially because in the incensement in reactionary energy utilized in unbridled and non-planning manner. Also, don not have proper planning to apply alleviate control measures is a chain in operation of air pollution. (3) Although, the colorful styles used to control air pollution to give accurate information and dependable data, the mechanism needed for similar assays were precious and cannot provide monitoring at high intensity situations across huge areas at different locales. Thus, the cast of air pollution is one of the safe and timely measures. (1) Figures of physical/ chemical monitoring tools, available for motoring air quality are piecemeal not only precious but much time consuming. Bio indicators in this environment are one of the stylish, a not expensive and natural agents, ratiocinating the presence of adulterants in air. Their response to any change in climate or pollution is important faster than any other biota. Change in growth as well as reduced or addition growth can be seen in comparison of lichen growth in weekend and healthy terrain. There are veritably important sensitive to air adulterants like $SO2$, $CO2$, CO etc.; thereby the number of lichen thalli in the weakened area is gradationally reduced and eventually it comes down to nil. For this reasons, the lichens are markedly absent in metropolises and artificial areas.

CONCLUSION

Finally, the project collected a few lines of data on lichen species while the lichens were successfully divided into three groups, lichen species proved a more difficult task with a large margin of error. However, the use of the lichen classification does have some effect on the air quality around the land; this becomes more accurate only when data is measured for pollution such as sulfur dioxide, carbon dioxide and nitrogen. The results of this important project can be used for future air quality projects especially when considering which lichen species show better air quality.Lichens serve as powerful and reliable bioindicators for detecting and monitoring air pollution. Their unique sensitivity to various pollutants, particularly sulfur dioxide, heavy metals, and nitrogen compounds, makes them invaluable for environmental assessment. Lichens are not only effective in revealing the presence of pollutants but also in providing insights into the long-term effects of air quality changes. Their response to pollution is often manifested through changes in their diversity, abundance, and physiological condition, making them a natural, non-invasive tool for environmental monitoring.

The use of lichens as indicators offers several advantages. They provide a cost-effective, continuous, and widespread method for monitoring air pollution over large geographic areas. Unlike technological instruments, which might be limited in scope or require frequent maintenance, lichens offer a more sustainable and accessible alternative, especially in remote or less-developed regions. Additionally, lichens accumulate pollutants over time, offering a historical perspective on air quality, which can be crucial for understanding trends and making informed decisions on environmental management.Moreover, the study of lichens contributes to our understanding of the ecological impact of air pollution. As organisms that are highly integrated into their ecosystems, changes in lichen populations can signal broader environmental shifts. This makes them not only indicators of pollution but also early warning systems for potential ecological degradation. The data gathered from lichen studies can inform policy decisions, guide industrial practices, and foster public awareness about the importance of maintaining clean air.However, while lichens are effective bioindicators, their use is not without limitations. Factors such as climate, habitat, and the presence of other pollutants can influence lichen health and complicate the interpretation of results. Therefore, lichen-based monitoring should be used in conjunction with other methods to provide a comprehensive picture of air quality.

REFERENCES

1. Aptroot A, Stapper NJ, Košuthová A, Cáceres ME. Lichens. [ed.] Trevor M. Letcher. *Climate Change (second edition)*. s.l.: Elsevier, 2016, pp. 295-307.
2. *"Lichen as a bio-indicator tool for assessment of climate and air pollution vulnerability."*. Kuldeep, Srivastava, and Bhattacharya Prodyut. 2015, Int. Res. J. Environ, pp. 107-117.
3. *Lichens redefined as complex ecosystems*. Hawksworth DL, Grube M. s.l.: new phytol, 2020, pp. 1281-1283.
4. Lücking, Robert, et al.. *"Fungal taxonomy and sequence-based nomenclature." Persoonia, 39, 1-11. DOI: 10.3767/persoonia.2017.39.01*. 2017.
5. Harris, R.C., & Lücking,. *"A new approach to the classification of lichen-forming fungi: the role of molecular phylogenetics." The Lichenologist, 50(4), 295-317. DOI: 10.1017/S0024282918000260*. 2018.
6. Hale, M. E., & Smith, C. *Lichens: An Illustrated Guide to the Fungi. Cambridge University Press. This book provides a comprehensive overview of lichens, including their role in environmental monitoring*. 2017.
7. 1. Cohen, J. S., & Dighton,. 1. *"Lichens as Bioindicators of Air Quality in Urban Areas: A Review." Environmental Monitoring and Assessment, 191(12), 728. doi:10.1007/s10661-019-7980-0. This review discusses how lichens are used to monitor air poll*. 2019.
8. 3. Kocourková, J., & Marušák,. *Lichen Diversity and Pollution in Forested Areas: A Study from Central Europe." Environmental Science and Pollution Research, 27(5), 5418-5429. doi:10.1007/s11356-019-07014-6. This study investigates the relations*. 2020.
9. 2. Zhao, L., & Liu,. 2. *"Response of Lichen Communities to Air Pollution: A Study in a Rapidly Urbanizing Area." Science of the Total Environment, 755, 142542. doi:10.1016/j.scitotenv.2020.142542. This paper explores the impact of urbanization and*. 2021.
10. Kumari B, Sharma GK, Vaish A, Kumar P, Ansari MJ. Conservation of Lichens. Chemistry, Biology and Pharmacology of Lichen. JUL 15, 2024, pp. 91-111.
11. *Assessing Ecological Risks from Atmospheric Deposition of Nitrogen and Sulfur to US Forests Using Epiphytic Macrolichens*. Geiser LH, Nelson PR, Jovan SE, Root HT, Clark CM. 2019, PubMed, pp. 1-87.
12. Asplund, J. and Wardle, D.A. *How lichens impact on terrestrial community and ecosystem properties*. s.l.: Biological reviews, 2017. 1720-1738.

13. Cannon, R. C., et al. (2017). *"Lichen Bioindicators of Air Quality: Recent Advances and Emerging Applications." Environmental Monitoring and Assessment, 189(8), 382. doi:10.1007/s10661-017-6030-1. 2017.*
14. López, M., et al. ". *"Lichen Diversity and Biomonitoring of Air Pollution in Urban Areas: A Case Study from Madrid, Spain." Science of the Total Environment, 621, 733-741. doi:10.1016/j.scitotenv.2017.11.274. 2018.*

CHAPTER 5

IMPORTANCE OF ACETIC ACID BACTERIA IN FOOD INDUSTRY

Divyansh Verma, Jyoti Prakash, Ruchi Yadav and Rachna Chaturvedi

Amity Institute of Biotechnology, Amity University Uttar Pradesh, Lucknow, 226028

Abstract

Acetic acid bacteria are a group of gram-negative bacteria, they have a shape from ellipsoidal to rod-shaped cells that have an obligate aerobic metabolism with oxygen as the terminal electron acceptor. Acetic acid bacteria are important microorganisms in the food industry as they can oxidise sugars and alcohols to organic acids and are the end products during the fermentation process. Acetic acid bacteria are also used in cellulose and sorbose production. Wine can be spoiled by the oxidizing ability of acetic acid bacteria. Acetic acid bacteria are very difficult in isolation, purification, identification and preservation. Acetic acid bacteria play a crucial role in the production of both conventional and industrial vinegar in the alcohol industry. They improve the taste and lengthen the shelf life of vinegar products. Furthermore, their contribution to the fermentation of kombucha, a well-known fermented tea, highlights their function in generating unique tastes and possible health advantages linked to probiotic activity. Although acetic acid bacteria have many benefits, they can also cause problems for the food business. This is especially true for wine production, where unintentional acetic acid production can cause spoiling. Thus, it is crucial to comprehend the ecology, metabolism, and regulation of Acetic acid bacteria to maximize their beneficial effects while reducing any potential drawbacks.

Keywords:
Acetic acid bacteria, food, fermentation, preservation.

INTRODUCTION

Acetic acid bacteria are aerobic, gram-negative, non-spore-forming. Acetic acid bacteria can have shapes from rod-shaped to ellipsoidal cells. Acetic acid bacteria can occur single, in pairs and chains. The size of acetic acid bacteria ranges from 0.4–1 μm wide to 0.8–4.5 μm long. For the growth of acetic acid bacteria, the optimum pH is 5–6.5 but they can also grow at low pH values between 3 and 4. Colonies of acetic acid bacteria can be grown by plating in solid culture media and they are selective for one strain or another. For isolation and phenotypic methods used for identification were done on different media. New techniques have been developed using molecular approaches, acetic acid bacteria were studied genotypically by DNA–rRNA hybridization, DNA–DNA hybridization and ribosomal RNA gene sequences (5S rRNA, 16S rRNA, and 23S rRNA). These techniques overcome the limitations of plating such as the inability to detect viable but non-cultivable (VBNC) cells and time requirements. Oxidizing the sugars and alcohols to organic acids is a well-known ability of acetic acid bacteria. The production of L-sorbose from D-sorbitol, and 2-keto-D-gluconate from D-glucose is done by genus *Gluconobacter*. Acetic acid bacteria are very important for the vinegar industry as these bacteria can produce high concentrations of acetic acid from ethanol. Production of sorbose from cellulose is also done with the help of acetic acid bacteria. Acetic acid bacteria also help in the formation of wine. [1]. Acetic acid bacteria were classified into two main genera: *Acetobacter* and *Gluconobacter in the early years* but today time twelve genera are recognized and accommodated to the family Acetobacteraceae, Alphaproteobacteria: *Acetobacter, Gluconobacter, Acidomonas, Asaia, Kozaki, Swaminathania, Saccharibacter, Neoasaia, Granulibacter, Tanticharoenia* and *Ameyamaea* [2].

Not only do acetic acid bacteria have practical uses, but they also play a significant part in food safety and spoiling. While their excessive activity can result in off-flavours and spoiling, understanding their metabolic pathways and growth conditions helps limit unwanted deterioration in wine and other alcoholic beverages. [3].

ACETIC ACID BACTERIA AND ALCOHOLIC BEVERAGES

Lambic Beer

Belgian lambic beers are very popular worldwide as they are refreshing, alcoholic beers have a fruity smell and they have little residual carbohydrates. They mature for three years after spontaneous fermentation of water, barley malt, and unmalted wheat. Due to the VBNC state of acetic acid bacteria cells, they are only sporadically recovered from the fermentation and maturation process of lambic beer. *Acetobacter lambici* and *Gluconobacter cerevisiae* are the two new species of acetic acid bacteria that have characteristics of acidic beer. [4].

Acetic acid bacteria together with the help of lactic acid produced by the lactic acid bacteria are responsible for the acidic flavour of lambic beer through the production of acetic acid. [5].

Acetic acid bacteria are essential to the traditional brewing of lambic beer because of the way they interact with wild yeasts and other microorganisms in the special environment of wooden barrels, adding to the richness of the beer. [6].

Water Kefir

Water kefir is low alcoholic beverage, it has acidic and fruity flavours, and it is a sparkling and refreshing beverage. By the fermentation of water, sucrose, fruits like figs and water kefir grains it is kept in a closed jar at room temperature for 2-4 days. Acetic acid bacteria are in low counts in water kefir fermentation, as revealed by both culture-dependent and culture-independent methods. Under aerobic conditions, there is growth leading to increased acetic acid concentrations which might be undesirable, and they remain in VBNC state under anaerobic conditions which means metabolically dormant and they start to grow when oxygen becomes available. [7]. Acetic acid bacteria contribute significantly to the nutritional and health benefits of water kefir by helping to produce bioactive substances. [8].

Kombucha

Kombucha is a very popular drink consumed usually in Asia. It is made from sweetened tea by fermenting, but some other plants like cereals and leaves, and animal raw materials like milk can be used and even mushrooms can also be used. It is a non-alcoholic beverage and has a specific flavour and acidity. The composition of microbes in kombucha depends on the origin, substrate and fermentation conditions. A wide range of yeast species can also be found in kombucha in addition of acetic acid bacteria.Acetic acid bacteria and yeast show a symbiotic relationship, hydrolysis of sucrose into glucose and fructose is done by yeast and it produces ethanol from glucose. With the help of glucose, fructose and ethanol acetic acid bacteria produce gluconic acid, glucuronic acid, acetic acid, D-saccharic acid-1,4-lactone and bacterial cellulose.Kombucha has multiple health benefits, it has properties like antimicrobial, antioxidant, and anti-inflammatory, and it also has anti-carcinogenic potential. [9]

Maintaining the balance of fermentation and guaranteeing the constant quality of the finished product depends on the symbiotic relationship between acetic acid bacteria and yeast in the kombucha culture, also known as SCOBY (Symbiotic Culture of Bacteria and Yeast) [10]. Figure 1 shows the Pathway of Kombucha formation.

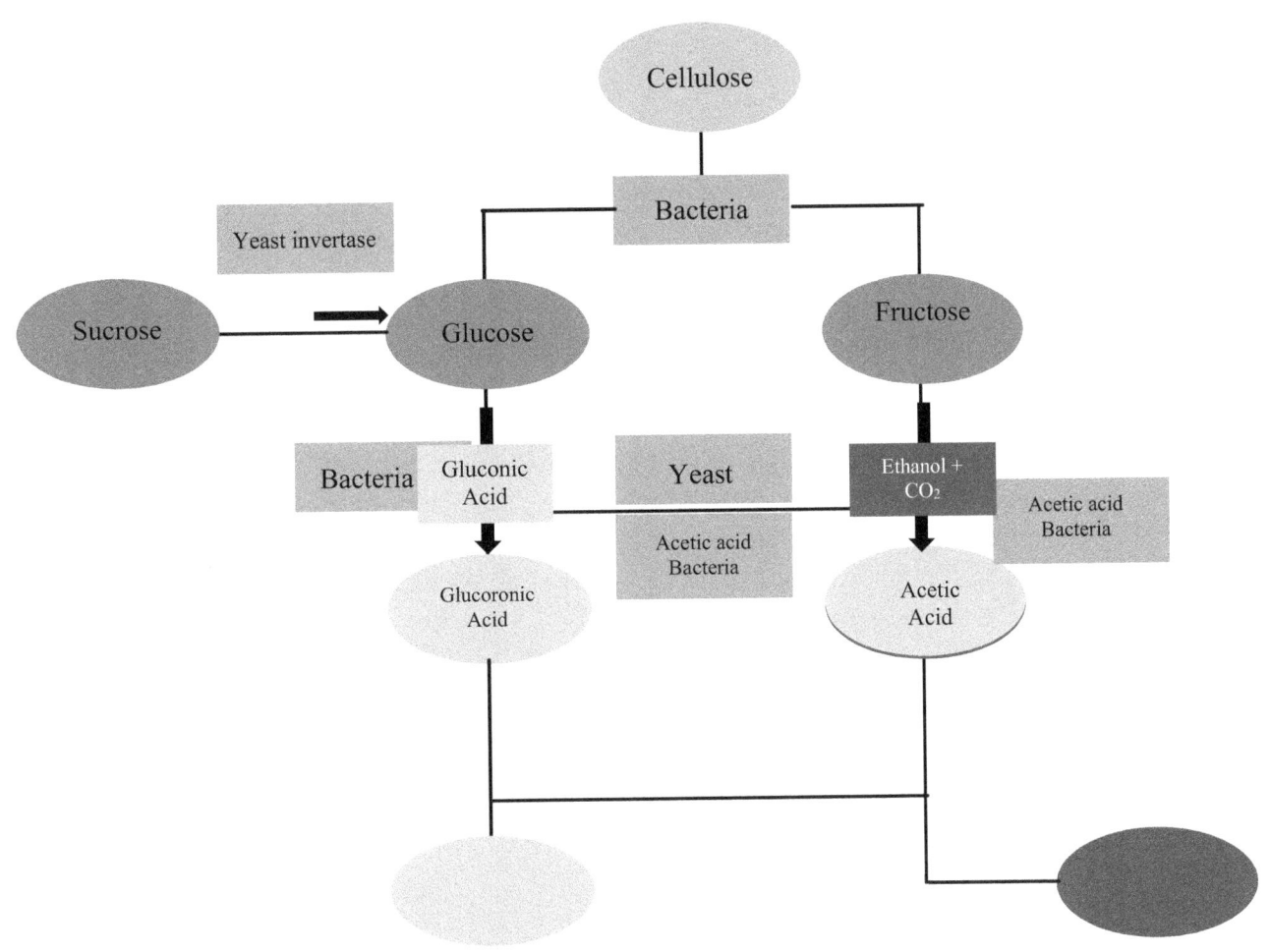

Figure 1: Pathway of Kombucha formation

ACETIC ACID BACTERIA AND VINEGAR PRODUCTION

All social classes all over the world consume vinegar despite not having a nutritional value. Vinegar has different raw materials, and manufacturing technologies and has a wide range of applications [11]. Vinegar is the result of double fermentation firstly alcoholic and then acetic acid. Mostly plant-originated substrate is used in the processing of vinegar.

Firstly, under anaerobic conditions *Saccharomyces cerevisiae* (yeasts) helps in converting fermentable sugars into ethanol, and then ethanol is converted into acetic acid by acetic acid bacteria due to the activity of two membrane-bound enzymes located on the outer surface of the periplasmic side of cytoplasmic membrane. Firstly, oxidation of ethanol to acetaldehyde takes place with the help of alcohol dehydrogenase and then acetaldehyde is oxidized to acetic acid by aldehyde dehydrogenase.

There are three main methods for the industrial production of vinegar –
1. Traditional Orleans or French which is a slow process it is done by surface acetification carried out in wooden barrel.
2. The fast generator which is done by production under forced aeration with wood chips or any other inert material.
3. The rapid submerged which is the modern or industrial process done by batch acetification with forced aeration and agitation.

Several genera of acetic acid production are included in the microbiota that leads to vinegar production which is a complex process. Acetobacter and Komagataeibacter are the species having a strong capacity to produce acetic acid but both genera have high resistance to high ethanol and acetic acid concentrations and they are very important characteristics required for industrial vinegar production. In vinegar production, thermotolerant acetic acid bacteria are introduced as they provide stable fermentation with lower cooling costs and particularly are advantageous in tropical countries. There are various applications of vinegar such as vinegar is used as a preservative, it is used as a flavouring agent, it is used as an ingredient in salad dressings, mustard, and mayonnaise. It can also be used as a cleaning agent and as a healthy drink in some countries [12]. Regular consumption of vinegar can have beneficial physiological health effects. Vinegar has therapeutic properties like antibacterial activity, regulation of blood pressure and glycaemia, antioxidant activity, prevention of obesity and prevention of cardiovascular diseases. Some types of vinegar have several bioactive compounds like polyphenols that contribute to taste. The efficiency of acetic acid bacteria in converting ethanol to acetic acid controls the vinegar's acidity, which is essential for its use in culinary and preservation applications, in addition to its quality and flavour profile [13]. Additionally, during vinegar fermentation, acetic acid bacteria help to produce useful chemicals and increase the bioavailability of specific nutrients, thus boosting the nutritional and health benefits of vinegar [14]

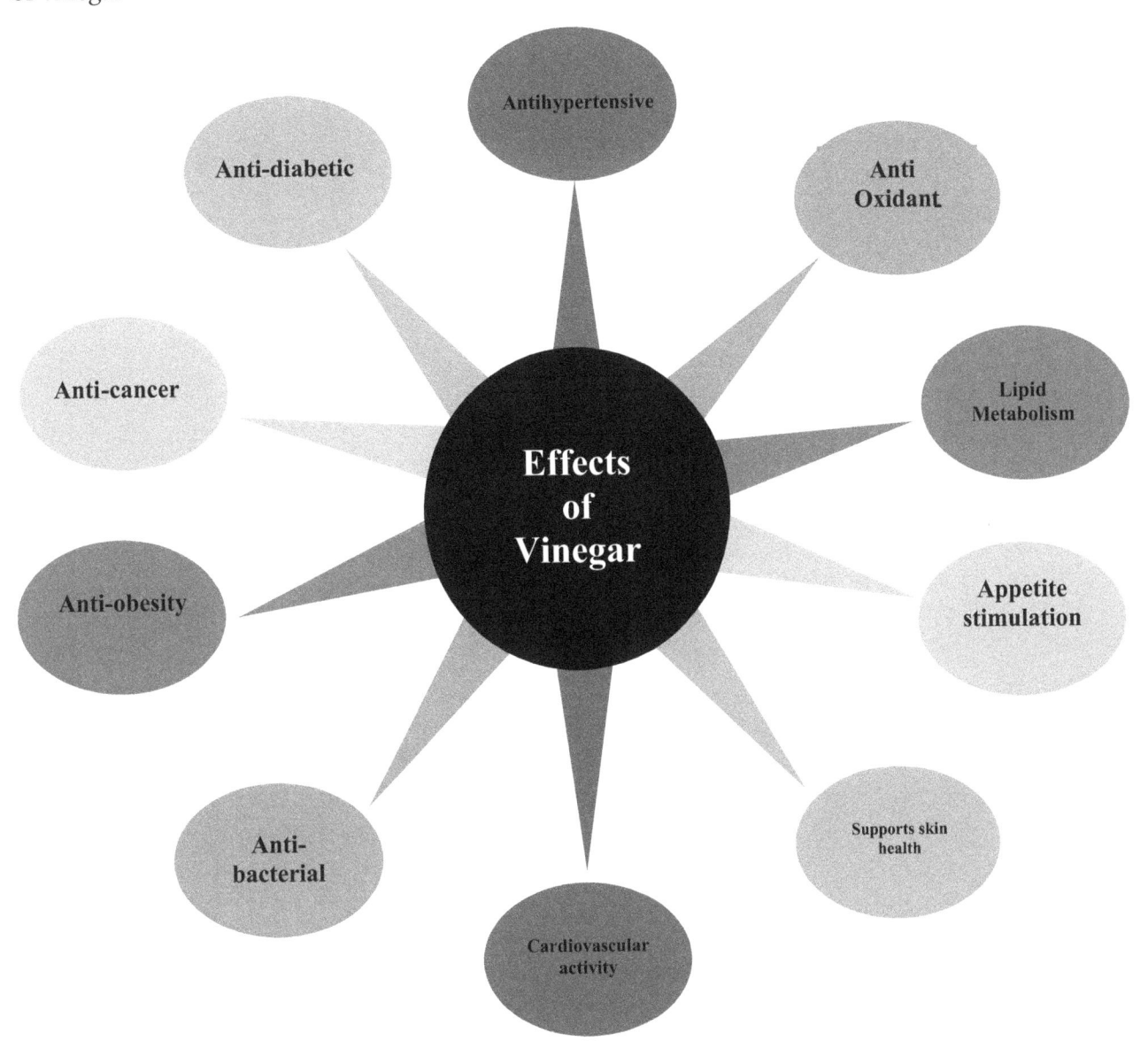

Figure 2: Effects of Vinega on human health

```
Raw material
    │
Upstream processes  ←  Yeast and Nutrients
    │
Alcohol Fermentation
    │
Store in storage tank
    │
Acetification
    │
Down Stream Processing  ←  Acetic Acid Bacteria
    │
Vinegar
    │
Marketing
```

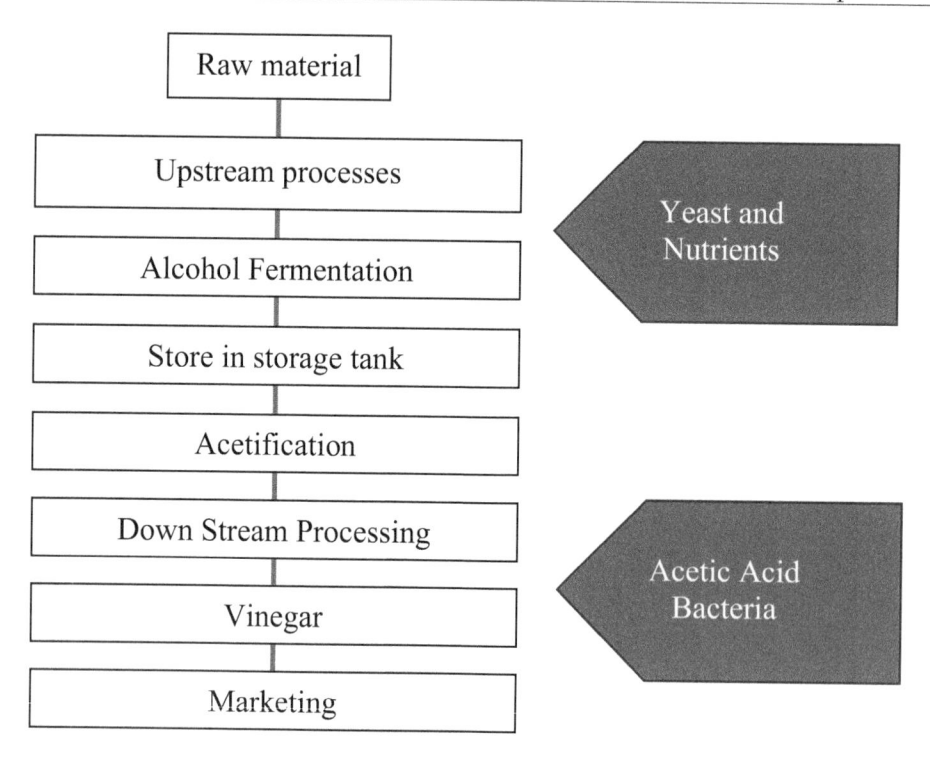

Figure 3: Production of Vinegar

COCOA AND ACETIC ACID BACTERIA

The main raw material for the manufacture of chocolate is cocoa bean (*Theobroma cacao L.*) [15]. Cocoa beans are taken from freshly prepared harvested cocoa pods. Yeast, Acetic acid bacteria, lactic acid bacteria and Bacillaceae regulate the complex fermentation process of cocoa beans [16]. There are three main stages of fermentation of cocoa beans –

In the first stage, the conversion of pulp sugars to ethanol and carbon dioxide via alcoholic fermentation is done by yeast simultaneously lactic acid bacteria species grow by converting glucose to lactic, acetic acid, ethanol and carbon dioxide.

In the second stage, there is an increase in lactic acid and a reduction of yeasts. In the third stage, acetic acid bacteria convert ethanol into acetic acid. Due to the increase in temperature and because of acetic acid and ethanol there is death of the seed embryo and that produces flavour, aroma and colour to the chocolate raw material.

The heat produced by acetic acid bacteria's exothermic reactions aids in killing the bean's germ, promoting adequate drying and averting unintended germination [17]. Since the final sensory qualities of the beans are directly influenced by their metabolic activities, acetic acid bacteria are necessary for the production of high-quality cocoa.

NATA DE COCO AND ACETIC ACID BACTERIA

In order to create nata de coco, a transparent, chewy treat created from coconut water, acetic acid bacteria (AAB) are essential. *Komagataeibacter xylinus*, formerly known as *Gluconacetobacter xylinus*, is the main species engaged in this process. It is well-known for its capacity to produce cellulose through oxidative fermentation. The bacterial cellulose that creates the gel-like matrix of nata de coco is produced when these microorganisms convert the glucose in coconut water into acetic acid. AAB's importance in this process goes beyond just producing cellulose; by controlling the fermentation conditions and guaranteeing the proper uniformity of the cellulose network, they also affect the texture and quality of nata de coco [18]. Recent research has emphasised the significance of acetic acid bacteria in the commercial production of nata de coco and the optimisation of fermentation parameters to improve yield and quality [19].

NATA DE PIÑA AND ACETIC ACID BACTERIA

The production of Nata de Piña, a typical Filipino delicacy prepared from pineapple juice, is greatly influenced by acetic acid bacteria (AAB). These microorganisms, especially those belonging to the Acetobacter genus, are in charge of turning the carbohydrates in pineapple juice into acetic acid, which helps to create the cellulose gel matrix that gives Nata de Piña its distinct texture. Similar to Nata de Coco, the production procedure involves AAB metabolising the juice's carbohydrates in an aerobic environment to produce microbial cellulose. The gel-like substance that is harvested and then transformed into the finished product is made of cellulose. Studies have indicated that several variables, including temperature, pH, and the initial sugar concentration, affect the growth and activity of the acetic acid bacteria involved in the manufacturing of Nata de Piña and affect its efficiency [20]. Furthermore, the strain of AAB employed might influence the quantity and calibre of nata generated; certain strains of *Acetobacter xylinum* are very efficient in this regard [21].

PALM WINE VINEGAR AND ACETIC ACID BACTERIA

Acetic acid bacteria (AAB) are essential to the production of palm wine vinegar, a traditional fermented beverage that is enjoyed all over the world. The primary byproduct of palm wine fermentation, ethanol, is oxidised by these bacteria, mostly belonging to the genera Acetobacter and Gluconobacter, to produce acetic acid, which gives vinegar its distinctively sour flavour. AAB's metabolic activity affects the product's flavour, fragrance, and general quality in addition to causing it to become more acidic. AAB activity can be increased and vinegar yield and quality can be improved by optimising fermentation conditions, such as temperature and oxygen availability, as recent research has shown [22]. Furthermore, creating starting cultures that could standardise and enhance the production of palm wine vinegar requires an understanding of the microbiological diversity of AAB in palm wine.In the face of industrialisation and modernisation, this biotechnological approach not only supports the preservation of traditional fermentation processes but also guarantees consistent product quality [23].

PICKLES AND ACETIC ACID BACTERIA

In order to improve flavour, texture, and preservation throughout the fermentation process, acetic acid bacteria (AAB) are essential to the creation of pickles. These bacteria, especially the Acetobacter and Gluconobacter species, convert the ethanol that lactic acid bacteria (LAB) make into acetic acid, which gives pickles their distinctively tangy flavour and serves as a preservative by decreasing pH and thwarting the growth of spoilage germs. The significance of AAB in conventional and industrial pickle fermentation methods has been brought to light by recent research. For example, a study showed that the regulated application of AAB improves the quality and stability of the finished pickled cucumber product [24]. Similarly, a study stressed that the development of pickles' sensory qualities depends on the interaction between LAB and AAB throughout the fermentation process. In order to maximise product quality and guarantee food safety in commercial pickle manufacturing, it is crucial to comprehend and control the activity of AAB in pickle fermentation [25].

FUTURE PROSPECTS OF ACETIC ACID BACTERIA

In recent times researchers are focusing on increasing yields and improving the quality of the product. New strategies are used for these purposes that are better acetification system, setting the optimum conditions for the processes, and selecting more productive strains. The improvement in tolerance to ethanol and acid would benefit not only vinegar production but also other bioprocesses. Greater yields and product quality will improve because of new substrates, measurement techniques, materials and optimization processes. To ensure the authenticity of the products and differentiate defective vinegars research has been focused on setting up validated analytical methods. New knowledge of genetics and the discovery of novel enzymes will further expand the potential use of acetic acid bacteria in biotransformations and other applications such as biosensors and alternative energy source. The food and pharmaceutical sectors use acetic acid bacteria-produced value-added products like sorbose and gluconic acid, which are produced through the biotransformation of different substrates. The efficiency and adaptability of acetic acid bacteria can be improved through the use of genetic engineering and biotechnology, which could open up new possibilities for the creation of functional foods and bio-preservatives. The introduction of acetic acid bacteria into food production systems contributes to the creation of novel goods and is consistent with the increasing customer preference for health-conscious and sustainable products.

CONCLUSION

New species and genera of acetic acid bacteria have been found in the last few decades. Their taxonomy and classification are based on molecular, physiological and biochemical characteristics. Acetic acid bacteria play an important role in food industries due to their brilliant ability to oxidise ethanol, sugar and sugar alcohols. Acetic acid bacteria help in the biosynthesis of pure and crystalline cellulose, a biopolymer with excellent industrial applicability. Acetic acid bacteria are used in production of beverages like kombucha, and lambic beer. Acetic acid bacteria help in the production of vinegar. Acetic acid bacteria also have antioxidant properties beneficial for human health.

REFERENCES

1. R. J. B. M. F. R. M. F. C.-G. R. J. H. &. S. W. A. Gomes, "Acetic Acid Bacteria in the Food Industry: Systematics, Characteristics, and Applications," Food Technology and Biotechnology, 56(2), pp. 139-151., 2018.
2. N. N. S. M. S. D. V. S. R. &. A. S. W. Yassunaka Hata, "Role of Acetic Acid Bacteria in Food and Beverages," Food Technology and Biotechnology, 61(1), pp. 85-103., 2023.
3. J. J. G.-G. I. S.-D. I. M. G.-M. T. &. M. J. C. Román-Camacho, "Latest Trends in Industrial Vinegar Production and the Role of Acetic Acid Bacteria: Classification, Metabolism, and Applications—A Comprehensive Review," Foods, 12(19), p. 3705., 2023.

4. F. W. A. D. J. M. A. M. D. H. M. &. V. L. A. Spitaels, "The Microbial Diversity of Traditional Spontaneously Fermented Lambic Beer," PLoS ONE, 10(4), p. e0133384., 2015.
5. D. V. L. De Roos J, " Acetic acid bacteria in fermented foods and beverages," Current opinion in biotechnology, pp. 1;49:115-9, 2018 Feb.
6. N. A. O. M. R. P. M. &. M. D. A. Bokulich, "Monitoring Seasonal Changes in Microbial Populations During Lambic Beer Fermentation," Applied and Environmental Microbiology, 83(23), pp. e00749-17., 2017.
7. S. M. S. D. V. S. R. A. S. W. Yassunaka Hata NN, "Role of Acetic Acid Bacteria in Food and Beverages," Food Technology and Biotechnology, pp. 25;61(1):85-103, 2023 April.
8. &. D. V. L. Laureys, "Microbial species diversity, community dynamics, and metabolite kinetics of water kefir fermentation," Applied and Environmental Microbiology, pp. 80(8), 2564-2572., 2014.
9. H. H. W. H. L. F. L. &. L. C. C. Liu, "The Isolation and Identification of Microbes from a Fermented Tea Beverage, Haipao, and Their Interactions during Haipao Fermentation.," Food Science and Technology Research, 24(4), pp. 711-718., 2018.
10. F. B. L. G. M. N. D. S. J. R. R. C.-M. J. L. B. S. &. D. G. Gaggìa, "Kombucha Beverage from Green, Black and Oolong Teas: A Comparative Study Looking at Microbial Community, Organic Acids and Volatile Compounds.," Beverages, 5(3), p. 29, 2019.
11. L. &. G. P. Solieri, "Acetic acid bacteria in traditional and industrial vinegar production: Molecular insights into their metabolism and influence on product quality," Trends in Food Science & Technology, 120, pp. 41-51., 2022.
12. Z. E. W. S. D. L. A. E. Lynch KM, "Physiology of acetic acid bacteria and their role in vinegar and fermented beverages," Comprehensive Reviews in Food Science and Food Safety, pp. 18(3):587-625, 2019 May.
13. B. &. P. A. Balasundaram, "Role of acetic acid bacteria in the industrial production of vinegar," Journal of Applied Microbiology, 121(2), pp. 302-309., 2016.
14. Y. Y. P. V. H. T. L. M. Y. &. N. S. Yamada, "Recent developments in the classification and role of acetic acid bacteria in vinegar production," International Journal of Food Microbiology, 336, p. 108914., 2021.
15. Z. V. G. D. B. K. V. P. &. D. V. L. Papalexandratou, "The key to acetate: metabolic fluxes of acetic acid bacteria under cocoa pulp fermentation-simulating conditions.," Journal of Industrial Microbiology and Biotechnology, 46(1), pp. 81-94., 2019.
16. G. V. M. M. J. A. I. &. S. R. F. Pereira, "Impact of Acetic Acid Bacteria on Cocoa Fermentation," Frontiers in Microbiology, 8, p. 116., 2017.
17. C. G. J. M. D. &. T. C. Kouame, "Towards a Starter Culture for Cocoa Fermentation by the Selection of Acetic Acid Bacteria," Fermentation, 7(1), p. 42., 2021.
18. M. R. e. a. Maria, "Optimization of nata de coco production using coconut water as the base medium," International Food Research Journal, 24(3), pp. 1073-1080., 2017.
19. S. M. Keshk, "Bacterial cellulose production and its industrial applications," Journal of Chemical Technology & Biotechnology, 91(11), pp. 2549-2560., 2016.
20. R. L. E. &. C. D. de Ory, "Effects of carbon source and growth conditions on the production of bacterial cellulose by Acetobacter xylinum," Biochemical Engineering Journal, 137, pp. 240-248., 2018.
21. G. L. S. E. P. G. L. &. A. A. G. Iñiguez-Covarrubias, " Industrial production of Nata: Strain selection and optimization of fermentation conditions," Biotechnology Advances, 35(5), pp. 707-719., 2017.

22. W. S. e. a. Tan, "Optimization of fermentation conditions for acetic acid production from palm wine," Journal of Applied Microbiology, 122(5), pp. 1234-1242., 2017.
23. e. a. Komagata, "Microbial diversity of acetic acid bacteria in palm wine fermentation and its impact on vinegar quality," Food Microbiology, 87, p. 103377., 2020.
24. Y. M. Y. C. X. &. H. H. Chen, "The impact of acetic acid bacteria on cucumber pickle fermentation," Journal of Food Science and Technology, 53(8), pp. 3204-3211., 2016.
25. K. S. K. &. K. S. Giri, "Interaction between lactic acid bacteria and acetic acid bacteria in the fermentation of pickles: A review," International Journal of Food Microbiology, 280, pp. 1-7., 2018.

CHAPTER **6**

IMPORTANCE OF LACTIC ACID BACTERIA IN FOOD INDUSTRY

Samridhi Mohan, Rachna Chaturvedi, Ruchi Yadav and Jyoti Prakash

Amity Institute of Biotechnology, Amity University Uttar Pradesh, Lucknow, 226028

Abstract

Lactic Acid Bacteria (LAB) have been crucial for the food industry due to their vast range of functional qualities and numerous applications. LAB are gram-positive and catalase-negative microorganisms that are utilized in making fermented food. These bacteria do not produce spores and have the morphological appearance of cocci or rods. LAB participates in metabolic activity by fermenting carbohydrates and creating lactic acid as their key end-product. Moreover, it also adds up the appealing taste qualities of fermented food. This metabolic activity also serves as a natural preservative by preventing the growth of spoilage and pathogenic microbes. Numerous traditional and specialty foods, such as yogurt, cheese, sauerkraut, sourdough bread, and pickles, which are cherished for their distinctive flavors, textures, and prolonged shelf life, are produced by LAB-driven fermentation processes. Several gastrointestinal issues, such as diarrhea, lactose intolerance, and irritable bowel syndrome, can be avoided and controlled through the use of LAB strains like Lactobacillus and Bifidobacterium. The antibacterial action that LAB is known to have against foodborne pathogens further makes the food products safe to consume. The current paper mainly focuses on the significance of LAB in the food industry and sheds light on their multifaced contributions to food production and food preservation.

Keywords:
Morphological, gastrointestinal, lactose intolerance, bowel syndrome

INTRODUCTION

Food Industry apparently utilizes lactic acid bacteria in many forms and facets. Their importance seems to be associated with the microbial actions of specific organisms, specifically those belonging to the group of lactic acid bacteria. Lactic acid bacteria (LAB) play a significant role in various fields such as food production, agriculture, and medicine. These bacteria can be described as Gram-positive, non-sporing, non-respiring organisms that exist in the form of cocci or rods. They primarily produce lactic acid when fermenting carbohydrates. [1]The core group of LAB is generally agreed to include four genera: Lactobacillus, Leuconostoc, Pediococcus, and Streptococcus. Additionally, there are several other genera that are currently classified within this group, namely Aerococcus, Alloiococcus, Carnobacterium, Dolosigranulum, Enterococcus, Globicatella, Lactococcus, Oenococcus, Tetragenococcus, Vagococcus, and Weissell. [2].Consequently, fermentative bacteria are commonly utilized as initial cultures in the food industry for industrial processing. Conversely, lactic acid can serve as a food additive in the edible products sector, even in the absence of lactic acid bacteria. This decision holds numerous potential uses. While this substance is primarily suggested for its role as an acidity regulator in various food products (including two optical isomers and their racemic mixture), there are specific considerations to be mindful of in certain situations regarding the maximum permitted quantities and the sole utilization of a single isomer. [3]

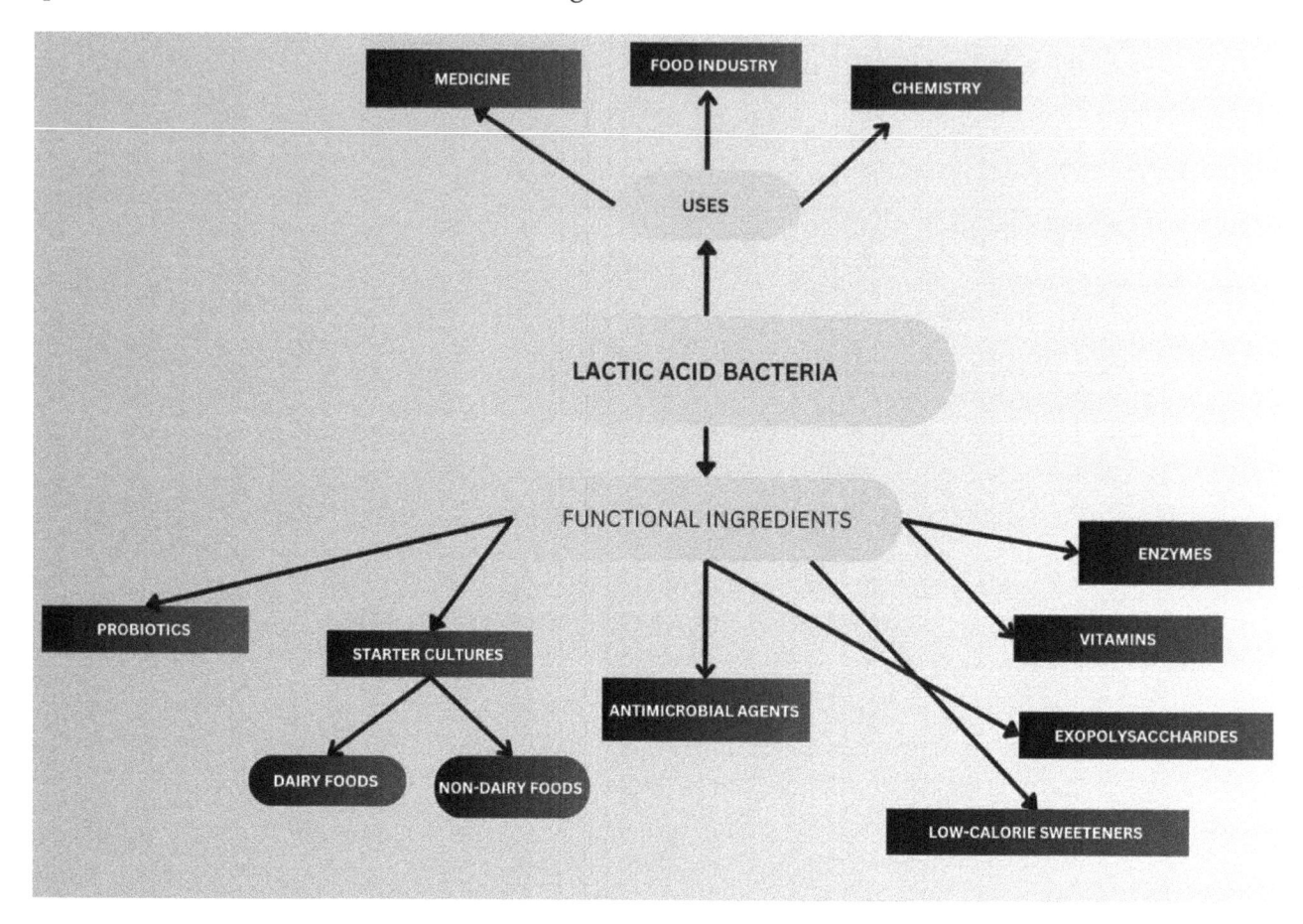

Figure 4: Role of lactic acid bacteria

One of the primary reasons for the extensive use of LAB in the food industry is their ability to inhibit the growth of spoilage and pathogenic microorganisms through the production of antimicrobial compounds. When ingested as a part of a balanced diet, LAB has gained popularity as probiotics because they provide a host of health advantages. The main processes responsible for generating taste in fermented food items mainly include three pathways: glycolysis, lipolysis, and proteolysis. [1] The primary by-product of the breakdown of carbohydrates is lactate, while some of the intermediate pyruvates can also be transformed into other compounds including diacetyl, acetoin, acetaldehyde, or acetic acid (some of which are crucial for the flavor of yogurt. Proteolysis plays a crucial role in the development of taste in fermented foods, while the impact of LAB on lipolysis is relatively insignificant. These constituents can undergo additional degradation, resulting in the production of various alcohols, acids, aldehydes, sulfur, and esters compounds that contribute to the creation of specific flavors in fermented food items. [4]

Lactic acid enhances the nutritional value of food, prolongs its shelf life, and contributes to its appealing flavor and texture. Bacteriocins are antimicrobial substances made of proteins, produced by specific LAB, and employed as food additives to hinder the growth of Gram-positive pathogenic and spoilage bacteria. [5] Probiotic lactic acid bacteria are beneficial for managing lactose intolerance, allergies, irritable bowel syndrome, diarrhea, and other related conditions. [6]

SIGNIFICANCE OF LAB IN THE FOOD INDUSTRY

Lactic acid bacteria, like microorganisms, seem to have occupied a vital role in the development of substances that benefit humanity. Throughout history, people have relied on them in households to produce yogurt, bread, and wine

These microorganisms are included in the food production process to add appealing color, flavor, aroma, and texture as well as improve the product's marketability. Similarly, In commercial fermentations, LAB is used to enhance flavor along with the texture of food and the feed as well. The importance of fermentation technologies in the food & beverage industry cannot be overstated because they allow for the preservation of edible goods, increase the shelf life of such products, and give them a desired sensory quality. [7]

Additionally, because probiotic microorganisms are present and the amount of nutrients in these products increases, they have a positive effect on the value of food in promoting health. They can also raise the level of microbiological safety. The metabolites created by the microorganisms involved in fermentation hinder the development of unwanted microorganisms and the formation of undesirable substances. Due to the potential reduction in the amount of chemical preservatives added to food, this occurrence is extremely desirable.

For the process of preservation and the production of healthy food items, LAB plays quite a significant role. Lactic acid fermentations are generally affordable and often do not necessitate substantial heat through the process of its preparation, making them efficient in terms of fuel consumption. Foods that have undergone lactic acid fermentation play a significant role in ensuring food availability for people across all continents. LAB play a crucial role in preserving and producing a wide range of food items, such as fermented fresh vegetables like sauerkraut and kimchi (a Korean preparation), pickled cucumbers, fermented cereal yogurts like Nigerian ogi and Kenyan uji, sourdough bread and similar wheat or rye-free products like Indian idli and Philippine puto, as well as fermented milk products like yogurts and cheese. During the procedure, lactic acid bacteria efficiently reduce the acidity level of the food, thereby restricting the development of other organism. Leuconostocs, lactic streptococci, and certain lactobacilli and pediococci are known to reduce the pH to a level between 4.0 and 4.5 before limiting their own growth. Although lactic acid bacteria generally thrive in various dietary substrates, they can be selective in artificial media.

Lactobacilli produce hydrogen peroxide and lactic acid by converting reduced nicotinamide adenine dinucleotide (NADH) through the activity of flavin nucleotide. This reaction occurs quickly when exposed to oxygen gas. This oxidation process enables lactobacilli to generate hydrogen peroxide and lactic acid. Moreover, glucose oxidase, a flavoprotein, also generates hydrogen peroxide, which possesses antibacterial properties against other microorganisms that cause food spoilage. Lactobacilli demonstrate a certain level of tolerance to hydrogen peroxide, which allows them to flourish. [7]

LACTIC ACID BACTERIA IN FOOD PRODUCTION

Starter Culture

A starter culture is a blend of microorganisms, comprising multiple cells of at least one microorganism, employed to expedite and control the fermentation of raw materials, resulting in the creation of fermented food. The group of LAB is actively involved in these procedures and has a well-established track record of being used in the secure manufacturing of fermented food and drinks. They rapidly lower the acidity level of the starting material by producing organic acids, predominantly lactic acid. Their capacity to generate various enzymes such as acetic acid, bacteriocins, aromatic compounds, ethanol, and exopolysaccharides carries considerable importance. This procedure improves the texture, shelf life, microbial safety, and sensory attractiveness of the final product. [8]. The process by which LAB ferments certain sugars to make fermented foods is lost to history. It is widely acknowledged that the majority of these food items belong to the dairy category, specifically cheese, yogurt, and fermented milk. Additionally, starter cultures are now utilized in the production of various cereal products, as well as fermented meat, fish, pickled vegetables, and olive products. [9] In the past, the production of these goods relied on natural fermentation, which occurred due to the development of microflora naturally found in the raw materials and their surroundings. These products were traditionally created through a process known as back slopping, where the resulting characteristics of the product depended on the dominance of the most well-adapted strains. Nowadays, the majority of fermented foods are made by incorporating carefully selected starter cultures that possess distinct and well-defined characteristics unique to each product. [9]

Non-starter Cultures

Non-starter LAB (NSLAB) are types of lactic acid bacteria found in cheese that are not part of the initial culture used in cheese production. These bacteria can be classified into four main groups: pediococci, mesophilic lactobacilli, enterococci, and leuconostoc. Every natural cheese analyzed so far has contained at least one of these four types of bacteria

These components are intentionally introduced during the production of fermented food, although their main purpose is not to produce acid. Instead, they serve to restore some of the biodiversity that is lost during pasteurization and improve hygiene. Non-starter LAB plays a crucial role in enhancing the flavor and expediting the maturation process.

Numerous categories of bacteria generate extracellular polysaccharides (EPSs), which may exist as capsular polysaccharides linked to the cell membrane or discharged into the adjacent environment. These substances play a important role in the manufacturing of dairy treats such as milk, cheese, fermented cream and yogurt. They enhance the overall quality of these products by improving their texture, mouthfeel, flavor experience, and shelf stability. Lactic Acid bacteria also involves in malolactic fermentation to produce wine, which is a secondary fermentation process where L-malate is converted to CO_2 and L-lactate via malate decarboxylase (malolactic enzyme). [10]

Bio-protective Cultures

Some specific laboratory experiments have demonstrated the ability of certain bacteria to produce bacteriocins. Bacteriocins are polypeptides synthesized by bacteria through their ribosomes, and they possess the ability to eliminate or impede the growth of other bacteria. Most of the antibacterial proteins known as bacteriocins, which have the ability to combat bacteria, are generally resistant to high temperatures. These bacteriocins are thought to function by binding to phosphate residues present on the cell membranes of target cells, resulting in the formation of pores and the activation of an enzyme called autolysin. This enzyme then facilitates the breakdown of the bacteria. Nisin, which is the most commonly acknowledged bacteriocin, finds extensive application in the food sector and is utilized as a food supplement in more than 50 nations. Its main usage is in processed cheese, dairy products, and preserved food items. [1]

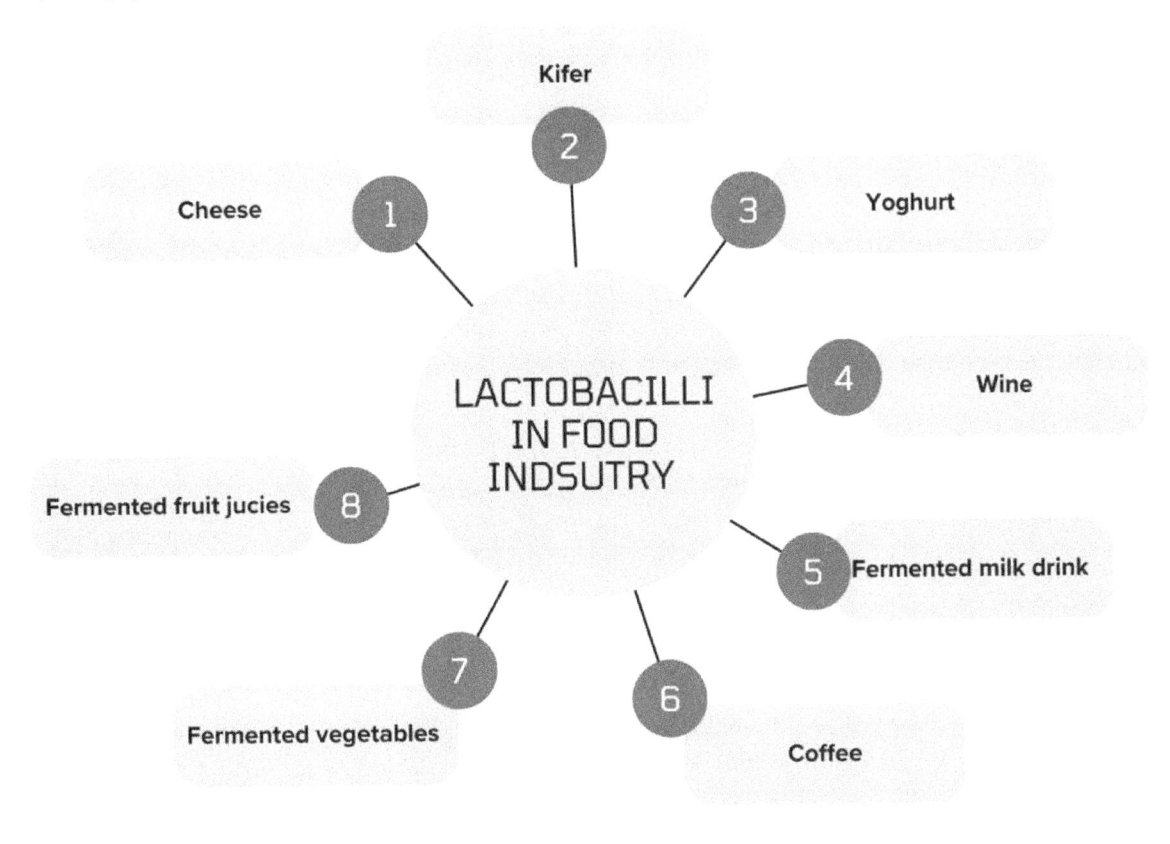

Figure 5: Lactic acid bacteria in food production

LAB IN PRESERVING AND ENSURING SAFETY OF FOOD

Research indicates that among the initial techniques used for preserving food, the fermentation of various food items with lactic acid bacteria stands out.. Some LAB strains have shown antimicrobial activity against pathogens that might cause food poisoning, including as bacteria, yeast, and filamentous fungus. Moreover, in recent times, a number of scientists have provided evidence showing that lactic acid bacteria possess the ability to counteract mycotoxins. [11] [12]

Lactic acid bacteria generate various substances such as lactic acid, organic acids, hydroperoxide, and bacteriocins, which are the primary factors contributing to their antimicrobial properties. Some studies propose additional antimicrobial mechanisms of lactic acid bacteria against infections, as well as their potential to counteract toxic by-products

These qualities are of utmost importance as there is a potential future where existing methods of chemical and physical preservation could be substituted by biological alternatives centered around lactic acid bacteria and their byproducts.

In the past, fermentation was carried out haphazardly, which led to low efficiency and inconsistent end-product quality. Currently, only a few starting cultures are employed in industrial production settings. However, regional and handcrafted goods frequently still rely on spontaneous fermentation. [11] [8]

LAB AGAINST FOODBORNE BACTERIAL PATHOGENS

A key issue in the food industry is the presence of foodborne pathogens during food production that cause a variety of illnesses linked to the intake of contaminated goods. The studies show that Lab inhibits the growth rate of foodborne pathogens such as *Salmonella spp* [1]*., Listeria monocytigenes* [13] *and Eshcerichia coli*.

The fermentation processes of these foodborne pathogens result in LAB in the production of a variety of metabolites with antimicrobial effects. These metabolites function by causing instability in the cell membrane, inhibiting the production of enzymes that create the cell wall, disturbing the balance of protons, and initiating the production of reactive oxygen species. As a consequence, this leads to increased levels of oxidative stress within the cell. [14]

According to the majority of scientific studies, the main strategy used to combat pathogenic microflora is to create conditions that make it impossible for them to grow, namely by lowering the pH of the environment to accommodate the significant amounts of lactic acid that they produce. The pH drop brought on by the presence of organic acids generated by LAB, such as acetic and propionic acid, effectively limits the growth of bacteria.

LAB AGAINST YEAST

Throughout history, yeasts have been regarded as the most influential microorganisms due to their integral role in the production of bread, alcoholic beverages, and dairy products. Moreover, yeasts have been extensively utilized as a valuable source of ethanol for fuel, as well as extracts, pigments, and biochemicals in the pharmaceutical industry.

However, yeast is involved in food and drink degradation. Because yeasts can thrive in low temperatures and pH levels and are resilient to physicochemical stress, they play a harmful function in biological processes. [15]

Yeasts can create unwanted microflora in the area where fermented items are produced. The production facility itself may be the source of yeast contamination in the food chain owing to an ineffective hygiene system that might encourage the creation of biofilm on technical surfaces. It is a problem connected to aerosols and overspray in cleaning. [4] Unwanted yeast can cause issues with product quality and safety in the food industry. Examples of such yeast include *Kloeckera apiculata, Brettanomyces bruxellensis, Rhodotorula mucilaginosa, Schizosaccharomyces pombe, Candida krusei, Candida parapsilosis, Debaryomyces hansenii, Pichia membranaefaciens,* and *Zygosaccharo* [4]

LAB AGAINST FILAMENTOUS FUNGI

The presence of filamentous fungi poses a notable problem that impacts both the food industry and agriculture as a whole. These fungi have a substantial impact on the economy due to their ability to contaminate food, animal feed, and cause agricultural diseases, resulting in significant economic losses. Additionally, they have the capacity to biosynthesize mycotoxins, harmful secondary metabolites.

Numerous writers have been able to offer proof that LAB activity has the effect of inhibiting the formation of filamentous fungus in fermented foods. The main reason why LAB activity is effective against the growth of filamentous fungi is due to the impact of their metabolites. These metabolites specifically target and weaken the cell membrane of the fungi, as well as hinder their ability to absorb amino acids. [16] The research demonstrated that isolated LAB strains obtained from traditionally fermented vegetables successfully hindered the growth of both toxicogenic and non-toxicogenic strains of filamentous fungus belonging to the Aspergillus, Fusarium, and Penicillium genera.

Table 1: Lactic acid bacteria strains against yeast and fungi in fermented prod

LAB Strains	Food	Inhibited Microogansims
1. *Lactobacillus harbinensis*	Yogurt	*Debaryomyces hansenii, Kluyveromyces lactis, Kluyveromyces marxianus, Penicillium brevicompactum,*
2. *Lactobacillus amylovorus*	Cheddar cheese	*Penicillium expansum* and environmental molds
3. *Lactiplantibacillus plantarum*	Wheat bread	environmental molds
4. *Lactobacillus amylovorus*	Sourdough	*Fusarium culmorum*
5. *Lactobacillus bulgaricus, L. plantarum*	Bread	*Aspergillus parasiticus* (only one tested in situ) and *Penicillium expansum*
6. *Lactiplantibacillus plantarum*	Citrus, apples and yogurt	*Penicillium roqueforti, Penicillium citrinum, Penicillium oxalicum, Aspergillus fumigatus, Aspergillus flavus*

CONCLUSION

In conclusion, the importance of lactic acid bacteria (LAB) in the food industry cannot be overstated. LAB plays a critical role in the creation, preservation, and improvement of a wide variety of food items. Their metabolic activities enhance the flavor, texture, and nutritional value of food. Additionally, they possess antibacterial properties that contribute to food safety by preventing the growth of harmful microorganisms. In addition to their direct impact on food production and human health, LAB also plays a vital role in preventing foodborne illnesses and managing food spoilage. By producing antimicrobial substances like organic acids, bacteriocins, and hydrogen peroxide, LAB ensures the safety and quality of food by inhibiting the proliferation of harmful bacteria that can cause illness and food deterioration. Overall, incorporating lactic acid bacteria into the food industry offers numerous benefits, including improved sensory characteristics, extended shelf life, enhanced nutritional value, and potential health advantages. As research in this field continues to advance, the use of LAB in food production is expected to increase, leading to the development of increasingly innovative and beneficial food products for consumers.

REFERENCES

1. E. P. B. A. a. A. A. Adetoye, "Characterization and anti-salmonella activities of lactic acid bacteria isolated from cattle faeces.," 2018..
2. W. B., " Lactic acid bacteria," benefits, selection criteria and probiotic potential in fermented food., p. 125, 2015. .
3. S. M. A. a. G. Caruso, "Spinger link," 25 may 2017. [Online]. Available: https://link.springer.com/book/10.1007/978-3-319-58146-0..
4. M. B. I. M. F. F. a. S. Z. G. Zara, "Yeast biofilm in food realms: Occurrence and control," Microbiol. Biotechnol., 2020.
5. C. A. O. A. O. O. M. P. Mokoena, "Applications of Lactic Acid Bacteria and Their Bacteriocins against Foodborne pathogen," 2021.
6. B. T, "Lactic acid bacteria: their applications in foods," 2018.
7. M.A, "A Review on Food Fermentation and the Biotechnology of Lactic Acid Bacteria," World J. Food Sci. Technol., 2018.
8. P. Z. D. Y. B. A. a. Z. F. W. Leonard, "Fermentation transforms the phenolic profiles and bioactivities of plant-based foods.," 2021..
9. A. B. T, "Dairy starter cultures.," In Papademas P, editor. Dairy microbiology, a practical approach., 2015.
10. Z... G. L. A. C. MS, "Linking wine lactic acid bacteria diversity with wine aroma and flavour.," 2017..
11. M. A. Admassie, "A Review on Food Fermentation and the Biotechnology of Lactic Acid Bacteria," 2018.
12. Y. S. Y. Z. P. Z. Z. M. a. X. Z. Z. Yu, "Potential use of ultrasound to promote fermentation, maturation, and properties of fermented foods," 2021.
13. M. C.-G. a. N. R. Miranda, "Expression of genes associated with stress conditions by Listeria monocytogenes in interaction with nisin producer Lactococcus lactis," 2018.
14. D. P. a. S. Kadyan, "Antifungal Lactic Acid Bacteria (LAB): Potential Use in Food Systems," in Novel Strategies to Improve Shelf-Life and Quality of Foods, Palm Bay, FL, USA,, 2020.
15. Y. W. H. B. M. M. a. L. S.. Matsubara, " Probiotic lactobacilli inhibit early stages of Candida albicans biofilm development by reducing their growth, cell adhesion, and filamentation.," 2016.
16. M. U. E. A. a. T. K. J. Awah, "T. Bio-preservative potential of lactic acid bacteria metabolites against fungal pathogens. Afr.," J. Microbiol. Res., 2018.

CHAPTER 7

HARNESSING BACTERIA FOR SUSTAINABLE CROP IMPROVEMENT

Palak Mishra, Jyoti Prakash, Rachna Chaturvedi and Ruchi Yadav

Amity Institute of Biotechnology, Amity University Uttar Pradesh, Lucknow, 226028

Abstract

A different branch that deals with the interaction of different soil types in a particular location is called soil microbiology. Because the soil contains many organisms, they also communicate with the soil to improve the quality and growth of plants. These organisms are bacteria, fungi, algae, protozoa etc. it could be. On the other hand, bacteria have been shown to be useful in improving crops since the early days of agriculture. Biological products such as biological fertilizers help increase fertility by providing essential nutrients in microbes added to the soil. Methanogens are different types of bacteria that produce mixed fuels. Plant growth promoting bacteria (PGPB) are particularly useful as stimulants of plant growth, nutrition and production. Rhizosphere bacteria, vegetation bacteria and endophytic bacteria play an important role in the direct and indirect biological control of plant growth and production. The main issue facing the globe today is population expansion and the enormous need for food. To increase output in agriculture, new techniques must be discovered and put into practice. Agricultural pesticides boost output, eliminate viruses, pests, and weeds, but they also severely damage the ecology. Microorganisms can be employed as a better alternative to agrochemicals due to growing concerns about their adverse effects on plants and microbial populations in the rhizosphere. The usage of plant growth-promoting rhizobacteria (PGPR) has been discovered to be a promising alternative to an outdated method that is commonly employed in agriculture but is increasing the burden of soil pollution. The rhizosphere's naturally existing soil microflora sticks to the surface of plant roots and has a positive impact on the plants [1].

Keywords:
microorganisms, bacteria, crops, pesticides.

INTRODUCTION

The use of microbes has been thought to benefit crops since Beijerinck discovered the nitrogen fixing ability of rhizobia in the early 1900s. However, the use of bacteria in plant development is limited because the mechanisms underlying plant-microbe interactions remain unclear. In recent years, the use of microorganisms to increase crop yield has gained renewed interest. Plant Growth Promoting Bacteria (PGPB) have been shown in research to support plant growth, nutrition and yield. PGPB is a biocontrol and plant growth regulator. They can also multiply microbial species in the soil, making nutrients easily accessible to plants. Many types of bacteria help crop such as rice and maize to be planted and grown in dry, rainy, windy weather. On the other hand, agricultural biology is the study of the growth, nutrition and yield of plants or crops. This research allows us to create and improve crops in many ways, such as genetic modification/crops that help us conserve natural resources and biodiversity. It also increases efficiency [2]. Plant growth promoting bacteria (PGPR), plant health promoting rhizobacteria (PHPR), Pseudomonas, Bacillus, etc. such as microbial bacteria. Help improve crops. Pseudomonas and Bacillus are plant growth promoters. PGPR is also known as biofertilizer as well as being used to improve soil quality and increase agricultural productivity. Plant diseases affect plant health and pose a threat to agriculture worldwide. The different defense mechanisms used by PGPRs allow plants to improve their health despite environmental stress. Microbes help crops maintain their original appearance, nutrition and more. Some commensal (Rhizobia, Bradyrhizobium, Mesorhizobium) and other non-symbiotic (Pseudomonas, Bacillus, Klebsiella, Azotobacter, Azospirillum, Azomonas) As a biological chemical it is used to transform the growth of living plants into many things.

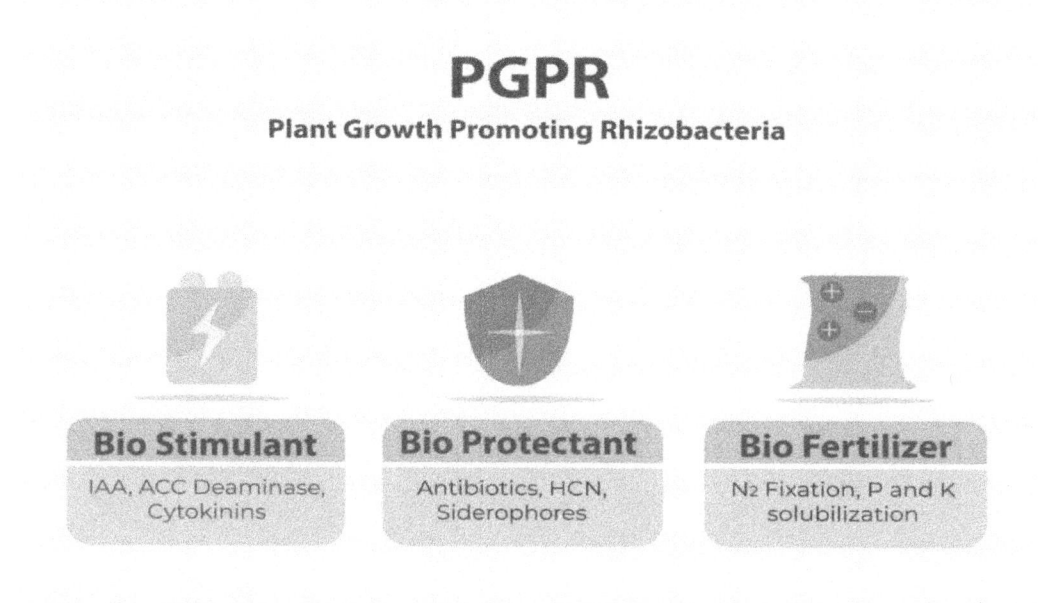

Figure 6: Role of PGPR in crop improvement. [3]

CROPS AND BACTERIAL ASSOCIATION

Leguminous crops like soybeans, peas, lentils, alfalfa, and clover form a symbiotic relationship with nitrogen-fixing bacteria called Rhizobium. These bacteria infect the factory's root nodes and convert atmospheric nitrogen into a shape that plants can use. This process enhances nitrogen vacuity in the soil, reducing the want for synthetic nitrogen diseases and promoting factory excrescency. Secondly, rice is one of the most important chief crops encyclopedically. Certain strains of bacteria, similar as Azospirillum and Burkholderia, have been shown off to enhance rice excrescency and nutrient uptake. These bacteria promote root evolution, boost nitrogen application effectiveness, and give defiance against pathogens and conditions [4]. For the potential to improve wheat growth and yield, bacteria including Azospirillum, Pseudomonas, and Bacillus have been investigated. They increase the availability of nutrients, foster the development of roots, and generate compounds that encourage plant growth, all of which have a favorable impact on wheat output. Investigations on the advantageous effects of bacteria on maize include Burkholderia and Azospirillum.

The increased root biomass, enhanced nutrient uptake, and production of chemicals that encourage plant growth caused by these bacteria increase agricultural productivity. Bacteria have recently received a lot of attention for their capacity to boost plant growth, nutrient uptake, and stress resistance. Bacteria like *Rhizobium*, *Azospirillum*, and Pseudomonas are well-known for their ability to promote plant growth through nitrogen fixation, phosphate solubilization, and the generation of phytohormones such as auxins and cytokinin's [5]. These bacteria improve agricultural yields while also increasing soil fertility, lowering the demand for chemical fertilizers. Recent research has demonstrated that bacterial inoculants can improve nutrient use, promote sustainable agriculture, and reduce the environmental effect of traditional farming practices [6].

BIOTECHNOLOGICAL ADVANCEMENTS IN BACTERIAL CROP IMPROVEMENT

Advances in biotechnology have made it possible to create genetically edited bacteria with improved crop enhancement capabilities. For example, microorganisms designed to express specific genes can impart disease resistance, improve tolerance to abiotic conditions such as drought and salinity, and increase crop nutritional content. Research undertaken after 2015 has shown that these genetically engineered bacteria have the potential to improve crop resilience and sustainability [7]. For example, Bacillus species modified to produce larger levels of stress-related proteins have demonstrated encouraging results in improving crop tolerance to severe environmental circumstances, so contributing to food security in the face of climate change. [8].

In the field of agricultural development, bacteria have emerged as a viable technique for increasing crop output, stress tolerance, and nutrient uptake. Plant Growth-Promoting Rhizobacteria (PGPR) play an important role in this process because they colonize the rhizosphere and facilitate nutrient acquisition, notably nitrogen, phosphorus, and iron, through mechanisms like as nitrogen fixation, phosphate solubilization, and siderophore synthesis. Additionally, several bacterial strains have been identified to produce systemic resistance in plants against diseases, lowering the need for chemical pesticides. For example, Bacillus subtilis and Pseudomonas fluorescens have been proven to boost resistance to numerous fungal diseases in crops such as wheat and rice. Furthermore, endophytic bacteria like Azospirillum species have been used into crop development projects because of their potential to promote root growth and improve water and nutrient uptake during droughts.

Recent advances have also concentrated on genetically modifying these bacterial strains to improve their efficiency and specificity, resulting in the generation of bioinoculants customized to specific crop requirements and environmental conditions. The incorporation of bacterial inoculants into sustainable agriculture techniques offers a promising avenue to increasing crop yields and resilience, especially in the face of climate change and declining soil fertility [9].

IMPORTANCE OF BACTERIA

Bacteria play a pivotal part in crop enhancement and agrarian practices. They form salutary symbiotic connections with shops, impacting their growth, health, and overall productivity. These relations have been exercised for colorful operations in ultramodern husbandry, leading to significant advancements in crop yield, nutrient uptake, complaint resistance, and environmental sustainability. One of the crucial ways bacteria contribute to crop enhancement is through nitrogen obsession. Nitrogen is an essential nutrient for factory growth, but atmospheric nitrogen is largely unapproachable to shops in its molecular form (N_2). Certain bacteria, known as nitrogen-fixing bacteria, have the capability to convert atmospheric nitrogen into a form that plants can use, similar as ammonia or nitrate. These bacteria form mutualistic connections with leguminous shops, forming technical structures called root nodes, where nitrogen obsession occurs. This process helps enrich the soil with available nitrogen, reducing the need for synthetic diseases and promoting sustainable husbandry.

Bacteria also play a pivotal part in promoting nutrient vacuity and uptake by shops. They enhance the solubility and vacuity of nutrients, similar as phosphorus, iron, and zinc, by releasing enzymes or organic acids that break down complex composites in the soil. This allows shops to pierce these essential nutrients more efficiently, leading to bettered growth and development. Likewise, bacteria can cover crops from conditions and pests. Certain salutary bacteria, known as biocontrol agents, produce antibiotics, enzymes, or poisons that inhibit the growth of pathogenic organisms. They can also contend with pathogens for coffers, populate factory shells, and stimulate the factory's vulnerable system, furnishing a defense medium against conditions.

By employing these salutary bacteria, growers can reduce the use of chemical fungicides and alleviate the negative impacts associated with them. Also, bacteria are involved in factory growth creation by producing factory growth-promoting substances similar as auxins, cytokinin's, and gibberellins. These hormones impact colorful aspects of factory growth, including root development, flowering, and regenerating. Also, bacteria can induce systemic resistance in shops, priming them to defend against unborn attacks by pathogens or pests. Bacteria are important in agricultural development because they promote plant growth and increase soil fertility. Beneficial bacteria, such as Rhizobium, Azospirillum, and Pseudomonas, help in nitrogen fixation, phosphate solubilization, and the generation of growth-promoting chemicals like indole acetic acid (IAA). These bacteria boost nutrient availability, root growth, and crop tolerance to environmental stressors. For example, nitrogen-fixing bacteria convert atmospheric nitrogen into a form that plants can absorb, minimizing the need for chemical fertilizers.

Furthermore, phosphate-solubilizing bacteria increase phosphorus availability to plants, which is essential for root development and general plant health [10]. Furthermore, bacteria are essential to biocontrol mechanisms, which reduce plant infections and pests, minimizing the need for chemical pesticides. Certain bacteria, such as Bacillus subtilis and Pseudomonas fluorescens, produce antibiotics, siderophores, and enzymes that prevent the growth of dangerous germs. This biological control not only preserves crops, but also promotes sustainable agriculture by preserving soil health and lowering environmental pollution. The use of bacteria in crop development is an environmentally benign method that improves crop production, quality, and resistance, making it an essential component of modern agricultural operations [11].

BIOLOGICAL NITROGEN FIXATION

Bacteria can be symbiotic ornon-symbiotic. Nitrogen is essential for the cellular conflation of enzymes, proteins, chlorophyll, DNA and RNA and is thus important in factory growth and in the product of food and feed. Nitrogen for tuberous legumes is handed by the symbiotic obsession of atmospheric nitrogen by nitrogenase enzymes in rhizobia. Rhizobium (formerly Agrobacterium), Frankia, Azospirillum, Azoarcus, Herbaspirillum, Cyanobacteria, Rhodobacter, Klebsiella, etc. are N- fixing bacteria which synthesize the unique nitrogenase enzyme responsible for nitrogen an obsession. Nitrogen- fixing bacteria in the soil souse the soil with inorganic nitrogen- containing composites, an important nutrient for crops.

When stable organisms die, the nitrogen in their biomass is released into the soil. In this way, they always increase soil fertility and enable growers to save on toxins. Symbionts are organisms that live in symbiosis with other organisms or with each other. For illustration, rhizobial bacteria that inhabit the root nodes of legumes give nitrogen- fixing exertion to these shops. Endosymbiosis and Ectosymbiosis are the types of symbiosis in soil. The symbiosis involved is good for non-nitrogen-fixing bacteria that follow grains and roots. This applies to rubrics like Azospirillum, Gluconobacter, Acetobacter, Herbacelia and Azospirillum.

Free- living N- fixing bacteria are also a source of N for crops. For illustration, rice directors add submarine Azolla ferns to their fields as green excreta, and Azolla serves as niche for Anabaena Azolla (cyanobacteria type), notorious for N- fixing parcels.

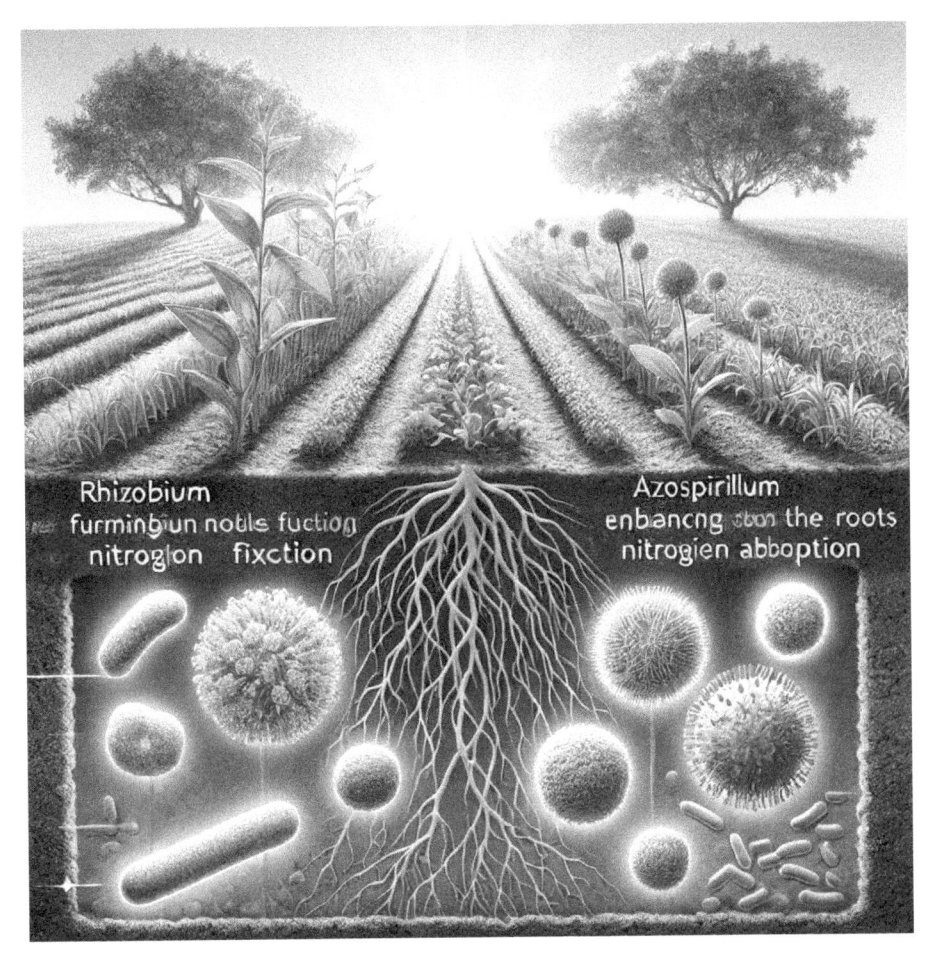

Figure 7: Representation of Rhizobium and Azospirillum nitrogen fixation.

AGRICULTURAL FORECASTING

Crop longevity will improve future quality of life, and agriculture is the essential process which could directly enhance the improvement of crops providing all the necessary requirements, growth and production of crops can be made viable just by ensuring presence of microorganisms which could improve the quality of crops. A sustainable strategy for both the environment and the economy will be the use of different species of bacteria promoting plant growth such as Rhizobium, Pseudomonas for fixation of nitrogen and numerous other species. Bacteria can affect agricultural forecasts in several ways, including nutrient management. The need for artificial fertilizers is reduced when bacteria, such as nitrogen-fixing bacteria, form symbiotic relationships with plants. These microbes convert airborne nitrogen into a form that plants can use, which consequently increases the production of crops. Furthermore, microorganisms support management and abiotic stress forecasting. Plant growth promoting rhizobacteria (PGPR) helps in withstanding the abiotic stresses like salinity, drought and severe ranges of temperatures. In addition to being utilized to enhance soil quality and increase agricultural output, PGPR are also referred to as biofertilizers. In general, nitrogen is regarded as one of the main nutrients that restricts plant growth.

Rhizobia in symbiosis with legumes are the most important nitrogen (N2)-fixing bacteria in agriculture, while other heterotrophic free living N2 fixers, cyanobacteria, and Frankia fix nitrogen in an amount slightly lesser than the Rhizobium species. Integrating bacterial inoculants into agricultural forecasting models has shown promise for long-term crop enhancement. Understanding the symbiotic relationships between bacteria and plants allows researchers to predict how these interactions would affect crop health under a variety of environmental stressors, such as drought or nutrient-deficient soils. For example, the incorporation of plant growth-promoting rhizobacteria (PGPR) in forecasting models has demonstrated tremendous potential for improving crop resilience and output, particularly in the face of climate change.

These innovations not only lead to more precise and effective agricultural operations, but they also minimize the need for chemical fertilizers and pesticides, resulting in a more sustainable agricultural system [12]. Agricultural forecasting, particularly in crop improvement, has grown more reliant on the usage of beneficial microorganisms. These bacteria play an important role in boosting crop resilience, growth, and yield by aiding nutrient uptake, generating growth-promoting chemicals, and providing disease resistance. Recent advancements in microbial genomics and precision agriculture have allowed for the identification and use of specific bacterial strains that may be adjusted to the needs of certain crops and environmental conditions. This approach, known as "bio-forecasting," enables farmers to predict the results of bacterial manipulations in crop systems, thereby optimizing inputs and increasing total agricultural productivity [13]. Recent advancements in data analytics, machine learning, and metagenomics have improved the accuracy of agricultural forecasts in this environment.

Large datasets from field trials, genomic sequencing, and environmental monitoring can be used to determine the most effective bacterial strains for various crops and environments. This strategy not only improves crop output estimates, but it also helps to generate more resilient crop types by utilizing intelligent bacterial inoculation strategies. Such improvements are critical in tackling climate change and food security issues because they enable more sustainable and efficient agriculture operations [14].

ENVIRONMENTAL CONSERVATION

For agriculture to be sustainable and crop output to increase, environmental preservation is essential. Utilizing the power of bacteria to boost crop development, often known as bacteria-mediated crop enhancement, is one cutting-edge strategy for achieving these objectives. This strategy supports environmental protection in addition to potential advantages for agricultural productivity. In many situations, including soil, bacteria are naturally occurring microbes. They have coexisted with plants for millions of years, developing complex connections that are essential for the health of ecosystems. Scientists have recently started to learn about some of the amazing properties of specific bacteria that help plants develop, make more nutrients available, and become more resistant to illnesses and environmental challenges. These helpful microorganisms are frequently referred to as bacteria that promote plant growth.

It is crucial to remember that, even though bacteria-mediated crop enhancement has a lot of potential, it should be used cautiously and in concert with other environmentally friendly agricultural methods. Thorough scientific investigation and field experiments must serve as the foundation for the choice and use of appropriate PGPB strains. To improve ecological harmony and reduce the danger of pathogen outbreaks or the emergence of bacterial resistance, additional techniques such as crop rotation, integrated Pest control, and diverse farming systems should be considered Finally, bacteria-mediated crop enhancement presents a promising route for agricultural environmental preservation. Farmers may cut back on chemical inputs, increase nutrient availability, boost stress tolerance, and improve soil health by utilizing the advantageous characteristics of bacteria that encourage plant growth. This strategy adheres to the fundamentals of sustainable agriculture, guaranteeing crop production's long-term profitability while reducing its environmental impact.

In the face of global environmental challenges, further research, innovation, and implementation of these practices can help build a more sustainable and robust agriculture system. Environmental conservation in crop improvement by bacteria focuses on using beneficial microbial communities to boost plant growth, resilience, and production while lowering dependency on chemical fertilizers and pesticides. Plant growth-promoting rhizobacteria (PGPR) are important in sustainable agriculture because they improve nutrient uptake, increase disease resistance, and promote plant development by producing phytohormones. These bacteria can fix atmospheric nitrogen, solubilize phosphorus, and create siderophores that sequester iron, thereby increasing plant nutrient availability. Farmers can reduce their use of synthetic inputs by adopting bacterial inoculants, hence minimizing agricultural practices' environmental impact.

This technique not only promotes soil health by preserving microbial variety, but it also helps to reduce the environmental implications of conventional agriculture, such as soil deterioration and water pollution. [15]. Furthermore, the employment of bacteria in crop development is consistent with the concepts of conservation agriculture, which stress minimal soil disturbance, soil cover, and crop rotation. Bacteria such as Bacillus, Pseudomonas, and Azospirillum are becoming more widely recognized for their role in improving crop resilience to abiotic stresses like drought and salinity, which are exacerbated by climate change. These bacteria contribute to natural resource conservation and food security under changing environmental conditions by boosting water use efficiency and stress tolerance. The incorporation of microbial solutions into crop enhancement strategies improves agricultural system sustainability while also promoting biodiversity and ecosystem services. Microbial-assisted crop enhancement is a viable approach to accomplishing long-term environmental conservation goals in agriculture.

RURAL DEVELOPMENT

Although a country's socioeconomic development is greatly influenced by rural development, agricultural advancement is a key component of rural development. Scientists and researchers have been looking into novel approaches to improve crop productivity and sustainability in recent years. Utilizing microorganisms to increase crop growth and yield and promote rural development is one potential strategy. The significance of bacteria in a few biological processes, such as the cycling of nutrients, the prevention and treatment of disease, and their symbiotic interactions with plants, has long been acknowledged. In the context of agricultural development, some helpful bacteria can form what is referred to as plant growth-promoting rhizobacteria (PGPR), a symbiotic relationship with plants. Plants with PGPR can increase their growth and general health by colonizing their root systems. Rural development in crop improvement by bacterial treatments has gained substantial traction in recent years, owing to the growing need to increase agricultural productivity in a sustainable manner. One of the most promising options is the introduction of plant growth-promoting bacteria (PGPB), which can improve crop productivity and resilience by increasing nutrient availability, disease resistance, and stress tolerance. Bacteria like Rhizobium, Azospirillum, and Pseudomonas have been extensively researched for their ability to fix atmospheric nitrogen, solubilize phosphates, and create phytohormones such as auxins and gibberellins. These microbial interactions are especially useful in rural settings where farmers frequently have limited access to artificial fertilizers and pesticides. Recent advances in microbial biotechnology have increased the use of bacteria in crop enhancement. Genome editing techniques, such as CRISPR-Cas9, are being used to create bacterial strains with improved characteristics, such as higher nitrogen fixation efficiency or resilience to environmental stress. Furthermore, the utilization of endophytic bacteria, which dwell within plant tissues without causing harm, is developing as a unique crop improvement strategy. These bacteria can help plants grow from within, providing protection against diseases and boosting nutrient intake [16]. The use of bacteria to improve crops in rural regions necessitates careful consideration of several variables, including the choice of the most appropriate bacterial strains, their compatibility with diverse crop species, and the optimization of application techniques. To further promote understanding of the advantages and appropriate application of bacterial crop enhancement approaches, education and literacy programme's should be put in place. The possibility for biological management of plant diseases is another advantage of utilizing microbes for crop enhancement in rural settings. Some PGPR are hostile to plant diseases, either through the production of antimicrobial chemicals or by the induction of systemic plant resistance. This can considerably lower disease occurrence and severity, improving crop quality and output. To sum up, using bacteria to improve crops has the potential to completely change rural development by raising agricultural production, cutting input costs, encouraging sustainable farming methods, and minimizing environmental effects. We can build a more resilient and affluent future for rural communities by incorporating bacteria-based crop improvement technologies into rural agriculture systems.

Harnessing Bacteria for Sustainable Crop Improvement   77

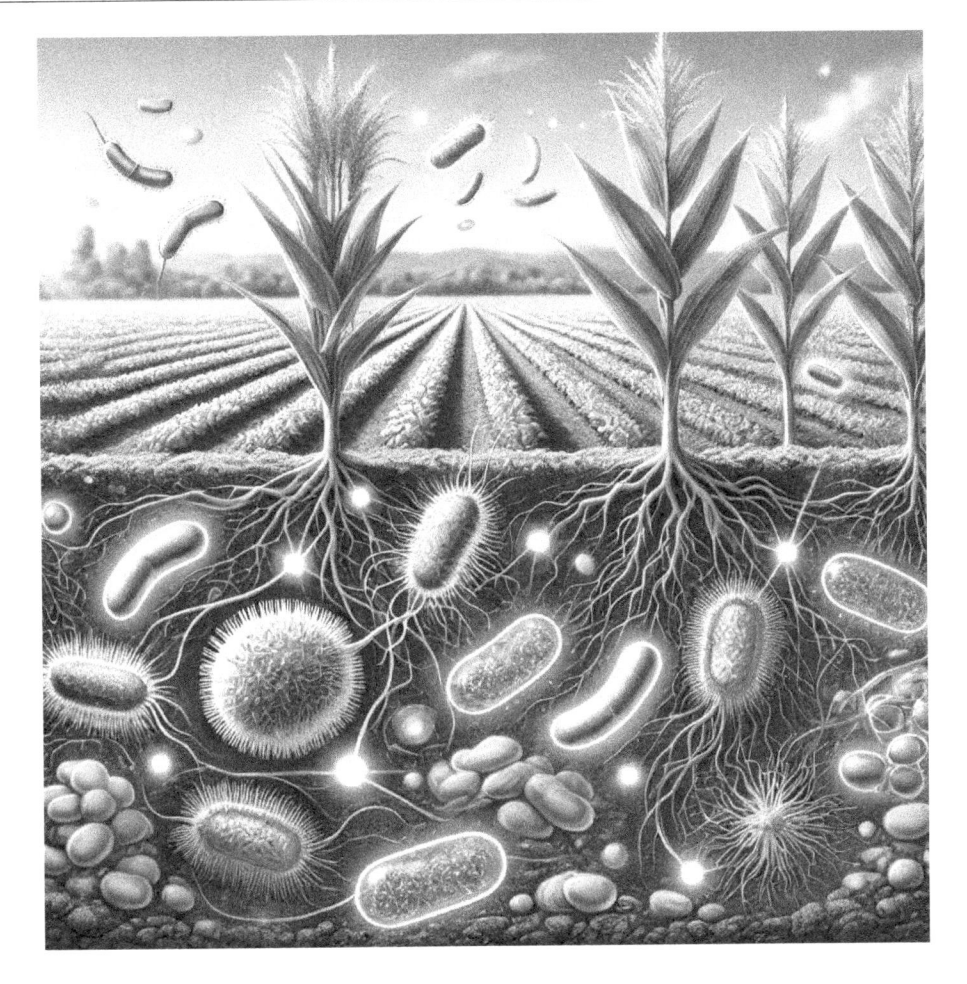

Figure 8: Diagramatic representation of improved quality of crops.

CONCLUSION

In conclusion, bacteria play a pivotal part in crop enhancement and have a significant impact on agrarian productivity and sustainability. Through colorful mechanisms, bacteria contribute to factory growth creation, nutrient cycling, complaint repression, and stress forbearance improvement. They establish salutary symbiotic connections with shops, enabling them to pierce essential nutrients and stimulating root development. Bacteria also produce bioactive composites and enzymes that cover shops against pathogens and pests. Also, bacteria are involved in nitrogen obsession, converting atmospheric nitrogen into forms usable by shops, thereby reducing the dependence on synthetic diseases and mollifying environmental pollution. Also, they prop in the solubilization and rallying of phosphorus, enhancing its vacuity to shops. Bacterial communities in the rhizosphere also grease the corruption of organic matter, contributing to soil fertility and nutrient cycling. Recent advancements in molecular ways and genomic studies have handed precious perceptivity into the complex relations between bacteria and shops, allowing for the identification and application of specific salutary bacteria for crop enhancement. These findings have paved the way for the development of biofertilizers and biopesticides that harness the eventuality of salutary bacteria to enhance agrarian practices. Likewise, the use of bacteria in crop enhancement offers sustainable and eco-friendly druthers to conventional agrarian styles. By reducing the reliance on chemical inputs, bacteria- grounded results contribute to the preservation of soil health, biodiversity, and overall ecosystem balance. They give a promising avenue for achieving advanced crop yields, perfecting food security, and mollifying the environmental impact of husbandry.

The use of microorganisms in crop development constitutes a huge step forward in sustainable agriculture. Recent research has demonstrated that beneficial bacteria, specifically plant growth-promoting rhizobacteria (PGPR), play an important role in increasing agricultural output and quality. These bacteria help plants absorb nutrients, boost root growth, and protect them from diseases. For example, Bacillus and Pseudomonas species have been intensively examined for their ability to create phytohormones, solubilize phosphorus, and fix atmospheric nitrogen, lowering the need for chemical fertilizers and increasing agricultural yield [17]. Bacterial engagement in crop enhancement also includes biocontrol. Beneficial bacteria operate as natural biopesticides by preventing the growth of dangerous pathogens and producing antibiotics, siderophores, and lytic enzymes. This not only decreases the need for synthetic pesticides, but also promotes a healthy ecosystem. The use of PGPR such as Pseudomonas fluorescens and Bacillus subtilis has been demonstrated to efficiently manage plant diseases in crops such as rice, wheat, and tomato, contributing to more sustainable farming methods [18].

The use of microorganisms in crop development is also being investigated using genetic engineering and synthetic biology. Scientists are creating strains with improved plant growth and resistance to stress conditions such as drought, salinity, and heavy metals by modifying bacterial genomes. For example, transgenic strains of Rhizobium and Azospirillum have demonstrated promise in enhancing nitrogen fixation and stress tolerance in leguminous and cereal crops, paving the door for more resilient agricultural systems in the face of climate change [19]. Furthermore, microorganisms have an important function in enhancing soil health, which is necessary for crop improvement. Soil microbiomes loaded with beneficial bacteria improve soil structure, water retention, and plant nutrient availability. The use of biofertilizers and bacterial inoculants is becoming increasingly important in organic farming for maintaining long-term soil fertility and sustainability. Studies have shown that using bacterial inoculants on a regular basis lead to the formation of beneficial microbial communities in the soil, which promotes robust plant growth and yields. Furthermore, microorganisms have a vital role in improving soil health, which is required for crop development. Soil microbiomes containing beneficial bacteria improve soil structure, water retention, and plant nutrient availability. Biofertilizers and bacterial inoculants are becoming increasingly significant in organic farming as a means of preserving soil fertility and sustainability.

According to research, employing bacterial inoculants on a regular basis causes the creation of beneficial microbial communities in the soil, which supports robust plant growth and yields. Finally, the use of bacteria in crop development provides a possible solution to current agriculture issues. As researchers continue to investigate the complex connections between plants and bacteria, it becomes clear that these microorganisms hold the key to constructing more sustainable and resilient agricultural systems. Advances in biotechnology and microbial ecology are anticipated to result in the discovery of new bacterial strains and uses, further changing crop improvement tactics. This strategy not only improves agricultural output and quality, but also adds to environmental sustainability by lowering dependency on chemical inputs.

REFERENCES

1. H. A. A. E. Radhakrishnan R, "Bacillus: A biological tool for crop improvement through biomolecular changes in adverse environments.," Frontiers in physiology, p. Sep 6;8:667, 2017.
2. M. K. M. S. S. S. P. C. R. D. P. K. P. Pallavi Mittal PM, "Plant growth promoting rhizobacteria (PGPR): mechanism, role in crop improvement and sustainable agriculture.," Inadvances in PGPR research, pp. pp.386-397, 2017.
3. "vectors," [Online]. Available: www.vecteezy.com/vector-art/11512354-pgpr-function-for-soil-the-role-and-working-principle-of-pgpr-plant-gowth-promoting-rhizobacteria.
4. D. N. S. S. K. J. G. S. V. A. Gupta K, "Plant growth promoting rhizobacteria (PGPR): current and future prospects in crop improvement.," Current trends in microbial biotechnology for sustainable agriculture., pp. 203-206, 2021.
5. Kumar, "Rhizobium and plant growth promoting rhizobacteria: implications in biocontrol of plant diseases.," Microbiological Research, pp. 169(1), 18-25, 2016.
6. R. Singh, "Plant growth-promoting bacteria: mechanisms and applications in agriculture.," Frontiers in microbiology, 2019.
7. M. Grover, "Plant growth-promoting rhizobacteria: Mechanisms and applications for crop improvement.," Plant Physiology and Biochemistry, pp. 158, 173-189, 2021.
8. Backer.R, "Plant growth-promoting rhizobacteria: Context, mechanisms of action, and roadmap to commercialization of biostimulants for sustainable agriculture.," Frontiers in plant science, pp. 9, 1473, 2018.
9. B. N. T. A. I. T. A. I. &. O. K. Raklami, "se of rhizobacteria to enhance the production of agricultural crops: benefits and challenges.," International Journal of Molecular Sciences,, pp. 20(19), 4683., 2019.
10. M. &. W. S. P. Kaushal, "Rhizobacterial-plant interactions: Strategies ensuring plant growth promotion under drought and salinity stress.," Agriculture, Ecosystems & Environment, pp. 231, 68-78. DOI:10.1016/j.agee.2016.06.031, 2016.
11. S. R. Y. A. G. D. D. S. P. K. &. P. K. D. Kumar, "Isolation and characterization of bacterial endophytes of Curcuma longa L. and their antagonistic activities against pathogenic fungi.," African Journal of Microbiology Research, pp. 10(32), 1270-1278. DOI:10.5897/AJMR2016.8034, 2016.
12. S. R. Y. A. G. D. D. S. P. K. &. P. K. D. Kumar, " Isolation and characterization of bacterial endophytes of Curcuma longa L. and their role in plant growth promotion.," International Journal of Environmental Science and Technology, pp. 13, 1903–1912. DOI: 10.1007/s13762-016-1014-2, 2016.
13. P. A. R. K. T. I. S. &. N. B. A. Vejan, "Role of Plant Growth Promoting Rhizobacteria in Agricultural Sustainability," A Review. Molecules, pp. 21(5), 573. DOI: 10.3390/molecules21050573., 2016.
14. J. K. &. B. W. D. Vessey, "The rhizosphere microbiome: Signaling and nutrient interactions.," Plant soil, pp. 417(1), 1-5. DOI: 10.1007/s11104-017-3237-9, 2017.
15. Y. d.-B. L. E. P. S. R. &. H. J. P. Bashan, "Advances in plant growth-promoting bacterial inoculant technology: formulations and practical perspectives," Plants and soil, pp. 378(1), 1-33, 2016.
16. S. M. W. M. S. R. Y. Y. H. A. S. &. L. I. J. Kang, " Isolation and characterization of a cadmium-resistant Enterobacter species from Alnus firma and its potential use in phytoremediation.," Plant and soil, pp. 416, 273-287, 2017.

A. S. &. N. V. Saharan, " Plant growth promoting rhizobacteria: A critical review," Life Sciences and Medicine Research, pp. 21, 1-30, 2017.

B. J. J. &. K. F. Lugtenberg, " Plant-growth-promoting rhizobacteria," Annual Review of Microbiology, pp. 63(1), 541-556, 2017.

17. R. K. L. &. D. S. Deshmukh, "Plant growth promoting rhizobacteria: An overview.," International Journal of Current Microbiology and Applied Sciences, pp. 5(4), 3-17, 2016.

CHAPTER **8**

ALGAL LIPID COMPONENT: A PROMISING TOOL IN BIOTECHNOLOGY

Vidushi Dubey, Vineet Awasthi, Gurjeet Kaur

Amity Institute of Biotechnology, Amity University Uttar Pradesh, Lucknow 226028

Abstract

Algae or thallophytes which are one of the most promising tool used by the scientists. These photosynthetic organisms have diverse ecology with variable size range and unique response to the subjected compounds. The algae could appear from microscopic to large seaweeds inhabiting fresh and salt waters including ponds, lakes, Warm Springs, rivers, oceans and in damp areas. With no limitation, they are also known to grow in snow, desserts and hard rocks. Though not confined to but majorly classified to: Blue green algae (cyanobacteria), Green alga (chlorophyta), Red algae (Rhodophyta), and Golden brown algae (Ochrophyta and Bacillariophyta). Their ecological origin makes them a boon for the biotech industry thereby utilising and exploiting their exceptional biological properties. Algae contain bioactive compounds including lipids, proteins, carbohydrates, pigments, vitamins, and minerals, nucleic acids, polyphenols and secondary metabolites. These bioactive components are enriching the scientific community with their substantial applications to mankind.

Keywords:
algal lipid, biotechnology, food supplement, biofuels

INTRODUCTION

Algae are rich in essential amino acids, proteins and peptides, good for fighting off bad bacteria, reducing inflammation and acting as antioxidants, for example phycocyanin isolated from spirulina is a well-known antioxidant. High on protein, it is a good food supplement and compliments the animal feed stock. It also adds economical input to the pharmaceutical industry. The carbohydrates including agar, alginate, carrageenan and simple sugars have potential health benefits. They contribute to boosting the immune system, combating viruses, biofuel production, gelling agent in food industry and has value addition in the pharma industry. Pigments present are chlorophylls, carotenoids, phycobiliproteins are known to carry antioxidant and anti-inflammatory properties. They are utilised as natural colours in cosmetics, food and are used as fluorescent markers by the scientists. Rich in vitamin A, B, C, D and E, they are used as natural supplements directly or indirectly. The polyphenolic compounds serve as a powerful antioxidant and reduce the risk of chronic diseases as well.

Among the biological compositions lipids are a rich energy storage hub. They are synthesized in the chloroplast by utilizing acetate in the acetate pathway and methylerythritol phosphate deoxy xylose phosphate pathway respectively producing fatty acids and terpenoids. The types of lipids are Fatty acids, Triglycerides, Phospholipids, sterols and glycolipids. The examples of fatty acids are Saturated, monounsaturated, polyunsaturated (PUFA) eg. Omega 3, omega -6, Eicosapentaenoic acid (EPA) and Docosahexaenoic acid (DHA). Adding to categorization lipids could be polar or non-polar. The non-polar lipids (no oxygen and few or no charges) produced by algae include waxes, fatty acids, glycerolipids and terpenoids. Polar lipids are oxygen rich and charged, including phospho and glycolipids forming the natural membrane components. Based on lipids being soluble in organic solvents they include fatty acids containing lipids (eg. glycolipids, betaine lipids, phospholipids and triglycerides), sterols (eg. cholesterol), hydrocarbons (example n-alkanes and alkenes), pigments (eg. chlorophylls, bacteriochlorophylls, and carotenoids) and toxins (eg. brevetoxins produced by the dinoflagellate *Karenia brevis*). The microalgal triglycerides serves as the primary storage form of energy and thus holds a promising exploration topic. It has been suggested that the cost of microalgae production should be brought down to grab the nerve of the market and post a competition to the existing market of the crude oil. The statistics of the market has been worked out. A group of scientists from Argentina characterized the algal homologs of diacylglycerol acyltransferase (DGAT) and phospholipid diacylglycerol acyltransferase (PDAT). Their observation led to novel clade DGAT 1- like proteins which were specific to the red alga and glaucophyta. DGAT which were exclusive to green algae had moderate similarity to chloroplast localised DGAT 3 which share the uncharacteristic proteins from the cyanobacteria. The *C. reinhardtii* DGAT 3 was found to increase the production of TAG thus contributing to biodiesel production. The biodiesel production has gained impetus over the past and so is the interest to enhance the algal biomass to meet the ends. Report suggests that thrantochytrids can produce over 50% of their weight in the form of triglycerides. To overcome the cumbersome downstream processing, a direct reaction of elemental sulphur with algal oil extract was conducted by Adarsh Gupta et al 2022. The unreactive oil obtained after environmental remedial measures as lithium sulphur battery cathodes, slow-release fertilizers and insulation was extracted for biodiesel production.

Micro and macro algal lipids serve as the major feed stock for the biodiesel production and Brazil tops the list as the largest biodiesel producer. Almost 50% lipid content in the form of triglycerides is a peculiar feature of microalgae. The triglycerides are converted to FAME (fatty acid methyl Ester) leading to biodiesel production is through transesterification process. Interestingly, Anesi et al showed obvious effect of temperature on the glycolipid fatty acid composition on freshwater dinoflagellates. A study conducted by wacker et al featured that, altered composition of fatty acids in some of the eukaryotic alga when subjected to radiations. This section clearly states that the demographic distribution of algae is a contributor to lipid content variability affecting the bioenergy production. The widespread human race with exploration as their favourite trait has welcomed algae as nutritional supplement complementing their regular food. Nonetheless, acceptance could be variable. Surprisingly, in the year 2012 an estimate showed 38% of the 23.8 million tonnes of seaweeds were directly consumed by the human population. In addition, agars, alginates and carrageenans also showed top up contribution to the data represented (FAO-2014).

Japan topping the list of nations consuming algae ranging from 9.6 to 11.0 g microalgae per day from the year 2010 to 2014. Laminaria species is used for preparing a food item called "**Kombu**", an important edible seaweed. In other countries monostroma is used to prepare "**Aonori**", porphyria is used for "**Asakusa nori**". Porphyra contains 35% protein 45% carbohydrate vitamin B and C and Niacin. No stock is used as a food material in South America. It requires special mention that the algae originated food is not confined to humans but also has been manifested as scientific diet for fishes prone to consuming Plecostomus and other algae like pure-cultured spirulina and chlorella algae. The nutrients are uniquely shaped for easy and long period consumption underwater.

Table 1: NUTRACEUTICALS AND FUNCTIONAL FOODS

S.No.	NAME	TYPE	CITATION
1.	*Spirulina platensis*	Microalgae	Ref. 1
2.	*Gelidium sp.*	Macroalgae (Red Algae)	Ref. 2
3.	*Palmaria palmata*	Macroalgae (Red Algae)	Ref. 3
4.	*Haematococcus pluvialis)*	Microalgae	Ref. 4
5.	*Undaria pinnatifida*	Macroalgae (Brown Algae)	Ref. 5
6.	*Tetraselmis sp.*	Microalgae	Ref. 6
7.	*Chlorella sp.*	Microalgae	Ref. 6
8.	*Spirulina sp.*	Microalgae	Ref. 6
9.	*Chaetoceros sp.*	Microalgae	Ref. 6
10.	*Cryptothecodinium sp.*	Microalgae	Ref. 6
11.	*Nannochloropsis sp.*	Microalgae	Ref. 6
12.	*Dunaliella sp.*	Microalgae	Ref. 6

The foods which generally contain algae are:
1. Dairy products as yogurt, whipped cream, chocolate milk, cottage cheese, ice cream and coffee creamers might contain carrageenan.
2. Non-dairy products might show presence of carrageenan and agar in almond milk, vegan cheese and non-dairy creamers.
3. Spirulina and chlorella powders are the seaweed supplements.
4. Sushi is prepared from "Nori" red algae a rich source of calcium and vitamin B.
5. Algae serving as spices for the preparation of seafoods. Kelp granules and dulse flakes can be sprinkled to add salty, umami flavour including iron and magnesium. Agar powder also adds to the list.

6. The quality of the meat products as pastry, frankfurters, sausages, fish and fish products to name a few are enhanced by adding algae.
7. The list of processed foods as cereal food products, baby food, ranch dressing, packaged Danishes and smoothies are enriched with carrageenan, alginate and Agar.

The questions are striving for answers related to the absorption, digestion and bioavailability of algae. A large population has acceptance stigma for consuming algae as food. The reports of discomfort after consumption of algae are available. The number of studies published are not enough to stem the eligible bioavailability. Revisiting the topic would give concrete food for thought of alcohol consumption. Counting on the variety of food products enriched by the algal species due to their rich lipid contents provides opening to industrial perspectives. The value addition to the food products has led to functional food targeting array of diseases under the nutraceutical domain. The bioactive compounds rich algal species have been studied rigorously to mellow down the deleterious effects observed during various alarming health concerns pertaining to cardiovascular disorders.

PHARMACEUTICAL

The bioactive compounds rich algal species have been studied rigorously to mellow down the deleterious effects observed during various alarming health issues pertaining to cardiovascular disorders neurodegenerative disorders immune disorders stroke and carcinomas. Algal lipids are explored for synthesising hormones, vaccines, growth factors, antibodies as well as immune regulators spirulina, chlorella, and Dunaliella commonly utilised in the pharmaceutical industry. For being rich source of W3, W6 fatty acids and range of vitamins as A,B,C,D,E And range of B1, B2, B5, B6 and B9, minerals, proteins and gamma linolenic acid. Reports of Haematococcus, Spirulina, Schizachyrium and Crypthecodinium producing eicosapentaenoic acid and docosahexaenoic acid Are known to play key role in the development of eyes and brain in the infants. Cyanobacteria serves as hydrating agents in food, pharmaceutical and cosmetic industry due to bulk production of exocellular polysaccharides. The immune system is triggered by the sulphated only secretaries isolated from the microalgae. *Chlamydomonas reinhardtii* has emerged to be very promising pharmaceutical biotechnological and RDT tool. It is a vaccine transporter. It is utilised for various antibodies vaccines, erythropoietin and VP 28 production. HbsAg and VP 28 and alpha galactosidase are produced in *D. salina*. Other promising micro algae used as RDT tool are chlorella, dunaliella and Senedesmus. Some examples of RT proteins are use as vaccines against swine fever virus, human papilloma virus. Toxins produced by Okhromonas species and *Prymnesium parvum* hold pharmaceutical relevance. A challenging approach was published as antiviral effects monogalactosyl diacyl glycerol (MGDG) extracted from Coccomyxa species. Antibactericidal properties of the algal lipids have been noticed. Both the differentially stained bacteria either gram positive or gram negative add ill fate when subjected to mixture of fatty acids from chlorella. The antibiofilm effects were noticed by the extracts of C. vulgaris and D. salina. The anticancerous effects of these micro and macro algae are no exceptions. antitumor effects against breast cancer lung cancer and human pro myeloid leukemia from the mono acetyl glycerides extracts of *Skeletonema malinois* has been published. The science revolving around nanoparticles has encompassed the biomedical word. The algal source of nanoparticle synthesis from alginate, carrageenan, laminarin and fucoidan type of polysaccharides interact with biomolecules due to hydrophilic groups present on the surface. The diatoms are rich source of silica nanoparticles along with DE (diatomaceous earth/frustules) used for gene delivery. Reports portray increased efficiency of transfection through receptor mediated endocytosis and decreased cytotoxicity. The porous silica diatoms were effective in hydrophobic and hydrophilic drugs example indomethacin and gentamicin. With modifications these diatoms were also used for core delivery of the drugs. With improved drug loading and controlled release. Many such algae are mounting the list as *Amphora subtropica, Coscinodiscus concinnus, Thalassiosira weissflogii.*

COSMETICS

Cosmetic word is coined from the Greek word "κοσμητικὴ τέχνη (*kosmetikē tekhnē*)". The literal meaning of the word is somewhat refrained in the interest of an individual to enhance their physical appearance. The global market of cosmetic industry is vast and has a major share in the economy of any country. The growing demand of the quality products with fancy packaging and attractive to the new genre. The cosmetic range unless stated otherwise comprises hygiene products beauty enhancing products and to some extent remedial measures pertaining to maintaining healthy looks the naturally occurring biochemicals have gained impetus due to their bioavailability and minimal side effects. Specifically, the plant based products have their sets of limitations being seasonal, demographical distribution, land availability and quality of the yield. To smoothen the skin, lipids serve as emollient and softening agent. The micro algal lipids comprise of triacylglycerols, phospholipids and phytosterol (glycolipids) which perform essential biological mechanisms to keep the skin soft and moistened. A complete range of algal metabolites are widely employed in products as moisturizers, sunscreens, antiwrinkle, antiaging and skin texture enhancing and treatment of hyperpigmentation. To uplift the economical reach of such products it is important to inflate the biomass production for industrial perspective.

Table 2: COSMETICS

S.No.	NAME	TYPE	CITATION
1.	*Haematococcus pluvialis*	Microalgae	Ref. 7
2.	*Laurencia sp.*	Macroalgae (Red Algae)	Ref. 8
3.	*Laminaria digitata*	Macroalgae (Brown Algae)	Ref. 9
4.	*Dunaliella salina*	Microalgae	Ref. 10
5.	*Ulva lactuca*	Macroalgae (Green Algae)	Ref. 11
6.	*Euglena gracilis*	Microalgae	Ref. 12
7.	*Enteromorpha sp.*	Macroalgae (Green Algae)	Ref. 13
8.	*Aphanizomenon flos-aquae*	Microalgae	Ref. 14
9.	*Hizikia fusiforme*	Macroalgae (Brown Algae)	Ref. 15
10.	*Codium fragile*	Macroalgae (Green Algae)	Ref. 16
11.	*Nannochloropsis oculata*	Microalgae	Ref. 17

Different aquatic microalgae mainly from the genera Aphanizomenon, Arthrospira, Chlorella, Desmodesmus, Dunaliella, Haematococcus, Nannochloropsis, Scenedesmus, and Spirulina are broadly used in cosmetic and cosmeceutical applications. Phytosterols are found in all microalgal species. In the recent past, *Diacronema lutheri, Tetraselmis sp. and Nannochloropsis sp.* have been identified as the highest phytosterol producers, but other studies on the phytosterols content are done in different microalgae species and classes. The various studies support the extraction of types of lipids from numerous microalgal species as *Nannochloropsis sp., Dunaliella sp., Schizochytrium sp., Isochrysis sp., Tetraselmis sp.* for polyunsaturated fatty acids, Sterols from *Diacronema lutheri, Tetraselmis sp., Nannochloropsis* sp., Waxes from *Euglena gracilis, Isochrysis* sp. and Carotenoids out of *Dunaliella salina, Haematococcus pluvialis, Chlorella sp., Scenedesmus sp., Muriellopsis sp., Spirulina sp., Porphyridium species.*

Table 3: BIOFUELS FROM ALGAE

S.No.	NAME	TYPE	CITATION
1.	*Oocystis submarina*	Microalgae (Green Algae)	Ref. 18
2.	*Chlorococcus infusionum*	Microalgae (Green Algae)	Ref. 19
3.	*Gracilaria salicornia*	Macroalgae (Red Algae)	Ref. 20
4.	*Chlamydomonas reinhardtii*	Microalgae (Green Algae)	Ref. 21
5.	*Schizochytrium sp.*	Microalgae (Stramenopiles)	Ref. 22
6.	*Sargassum sp.*	Macroalgae (Brown Algae)	Ref. 23
7.	*Parachlorella kessleri*	Microalgae (Green Algae)	Ref. 24
	Dunaliella tertiolecta	Microalgae (Green Algae)	Ref. 25
9.	*Laminaria japonica*	Macroalgae (Brown Algae)	Ref. 26

Genera of microalgae used for biofuel production are Chlamydomonas, Chlorella, Nannochloropsis, Synechocystis, Tetraselmis, Monoraphidium, Ostreococcus, Tisochrysis, and Phaeodactylum. TAGs from vegetable oils are also used in the formulation of lubricants available in the market.

Table 4: BIOGASOLINE AND JET FUELS

S.No.	NAME	TYPE	CITATION
1.	*Aurantiochytrium limacinum*	Microalgae (Stramenopiles)	Ref. 27
2.	*Thalassiosira pseudonana*	Microalgae (Diatoms)	Ref. 28
3.	*Tetraselmis suecica*	Microalgae (Green Algae)	Ref. 29
4.	*Isochrysis sp.*	Microalgae (Haptophytes)	Ref. 30
5.	*Acutodesmus obliquus*	Microalgae (Green Algae)	Ref. 31

The reign of fossil fuels has powered progress for generations, but the environmental toll is undeniable. Climate change, air pollution, and the looming depletion of these resources necessitate a swift transition towards a sustainable future. The quest for alternate source of fuel has been recorded and studied time to time in human history. A prominent example of this can be seen in the deadliest times of second World War. Two European scientists Richard Harder and Hans von Witsch suggested cultivating diatoms, a type of algae to produce food and biofuel. During 1970s the Unites States government led research on potential of algae as fuel but this couldn't progress further and was put to halt in the year 1990s due to technological limitations and economic considerations.

Biofuels, a revolutionary class of fuels derived from renewable organic materials, offer a beacon of hope in this critical juncture. The evolution of biofuels has seen a shift towards more sustainable options. First-generation biofuels, derived from food crops, raised concerns about land use. Second-generation options, like waste-based fuels, addressed this but still require dedicated land. This is where the paradigm shifted to algae. As third-generation biofuels, these fast-growing organisms thrive in non-arable land and even wastewater. Remarkably, algae produce oil-rich biomass, a perfect feedstock for sustainable biofuels like biodiesel and jet fuel. Their minimal land footprint and rapid growth make them a promising solution for the future of clean transportation fuels. This chapter explores the potential of algae and the various biofuels they can generate, paving the way for a greener transportation landscape. Algal lipid biofuels offer a promising avenue for transitioning away from fossil fuels. This process holds advantages over traditional plant-based biofuels. Algae are land independent. Unlike crops that require vast tracts of arable land, algae can thrive in non-arable land and even saltwater environments. This eliminates competition with food production and allows for cultivation on marginal lands unsuitable for agriculture. Secondly, algae function as natural wastewater cleaners. They can utilize wastewater for growth, mitigating pollution while simultaneously producing biofuel feedstock. Finally, algae are voracious consumers of carbon dioxide (CO_2). As they grow, they capture atmospheric CO_2, a major greenhouse gas, contributing to environmental sustainability.

The potential benefits of algal biofuels extend beyond just their production process. Research and development have yielded a diverse range of biofuels derived from algae, including familiar options like bioethanol and bio gasoline, alongside more specialized fuels such as green diesel, jet fuel, and biobutanol. This versatility ensures that algal biofuels can be tailored to meet a variety of transportation and industrial fuel needs. But over the years of extensive research the green renewable energy industry has shifted its focus on the type of biofuel which is been produced from the microalga lipid biofuel. Microalga lipid biofuel refers to biofuel derived from the lipids (oils and fats) accumulated by microscopic algae. These single-celled organisms, despite lacking the intricate structures of plants, are attracting significant interest as a potential alternative source for biofuels. There are several key reasons behind this interest include, microalgae are capable of accumulating high levels of lipids within their cells. In some species, this lipid content can reach up to 70% of their dry weight, far exceeding the oil content of traditional energy crops. Secondly, microalgae boast impressive growth rates. Compared to terrestrial plants, they can multiply much faster, allowing for quicker biomass production for biofuel feedstock. Decisively, microalgae excel in photosynthesis, efficiently converting sunlight and carbon dioxide into energy-rich organic compounds, including lipids. This photosynthetic prowess surpasses the complex mechanisms employed by traditional plants, offering a potentially more efficient biofuel production system. A diverse range of algal biofuels are currently under investigation and development.

Table 5: ALGAL FUELS

FUEL TYPE	DESCRIPTION	ADVANTAGES	DISADVANTAGES	Generation
Biodiesel	Produced from the fatty acids (lipids) of algae through transesterification.	High energy density, compatible with existing diesel engines	High production costs, requires efficient oil extraction	Second generation
Bioethanol	Produced through fermentation of algal sugars.	Renewable, biodegradable, can be blended with gasoline	Lower energy density than gasoline, complex fermentation process	Third generation
Biobutanol	A longer-chain alcohol fuel produced through fermentation of algal sugars.	Higher energy density than ethanol, less volatile and corrosive	Requires specific strains for production, research and development stage	Third generation
Biogasoline	Synthetic gasoline produced using algal biomass as a feedstock.	High energy density, can be directly used in existing engines	Complex and energy-intensive production process, technology under development	Third generation
Methane	Produced through anaerobic digestion of algal biomass.	Versatile fuel source, existing infrastructure can be utilized	Lower energy density than other options, requires additional processing	Third generation

Bio-aviation fuel has emerged as a critical component of sustainable aviation due to the aviation industry's substantial contribution to greenhouse gas emissions. Traditional jet fuel significantly impacts both local air quality and global climate patterns. Biofuels offer a promising alternative with the potential to substantially reduce emissions. Furthermore, relying solely on fossil fuels exposes the aviation sector to fuel price volatility and geopolitical risks. Biofuels provide a means to diversify fuel sources and enhance energy security. Additionally, they offer economic benefits such as job creation and reduced reliance on fuel imports. With ongoing advancements in biofuel production technology, their integration into the aviation industry is becoming increasingly feasible. The aviation industry faces a pressing challenge: mitigating its environmental impact while facilitating global travel. Sustainable aviation fuels (SAFs) are the key to unlocking a cleaner future for air travel. In this quest for sustainable alternatives, algal lipids emerge as a frontrunner. Microscopic algae, with their inherent ability to produce oil-rich compounds, offer a compelling proposition for SAF production. The systematic advantages of algal lipids are threefold. Firstly, their high energy density translates to fuels with a high energy content per unit volume, a critical factor for powering long-haul flights. Secondly, algae boast rapid growth potential, enabling the sustainable and potentially large-scale production necessary to meet the ever-increasing demand for clean aviation fuels. These advantages are further bolstered by the inherent versatility of algal lipids. While ideal for biofuels, they also hold promise for other applications, promoting the economic viability of large-scale cultivation. Therefore, algal lipids, with their high energy density, rapid growth potential, and economic viability, present a promising solution for the development of sustainable aviation fuels.

Although algal biofuels offer a sustainable alternative, significant challenges persist. Large-scale cultivation requires substantial infrastructure, impacting ecosystems and incurring high production costs. While some algae strains boast impressive oil content, the average yield is currently low, necessitating more algae to produce a desired fuel amount, further straining efficiency. Additionally, large-scale freshwater use for open pond systems poses a competition for water resources.

Fortunately, research into alternative water sources like saltwater and wastewater, coupled with advancements in cultivation techniques, harvesting methods, and strain optimization, offer promising avenues for overcoming these hurdles and unlocking the full potential of algal biofuels. While significant hurdles exist in the large-scale production of algal biofuels, the potential for a clean and sustainable transportation future fuelled by algae remains undimmed. Ongoing research in cultivation techniques, strain optimization, and efficient harvesting methods offers exciting possibilities to overcome these challenges. As these advancements progress, algal biofuels have the potential to become a cost-effective and environmentally responsible alternative to traditional fossil fuels. Continued investment and innovation can transform algae cultivation into a cornerstone of a cleaner and more sustainable transportation system, leaving a lasting positive impact on our planet for generations to come.

The aviation industry stands at a critical juncture, compelled to reduce its environmental impact while maintaining global connectivity. **Bio-aviation fuel has emerged as a critical component of sustainable aviation due to the aviation industry's substantial contribution to greenhouse gas emissions.** Sustainable aviation fuels (SAFs) are essential for achieving these goals. Among the potential feedstocks for SAFs, algal lipids offer a compelling proposition. These oil-rich compounds produced by microscopic algae possess the potential to deliver high lipid yields, carbon neutrality, and renewability, making them a promising resource for aviation fuel production. Traditional jet fuel significantly impacts both local air quality and global climate patterns. Alternatively, biofuels offer a promising alternative with the potential to substantially reduce emissions. Furthermore, relying solely on fossil fuels exposes the aviation sector to fuel price volatility and geopolitical risks. Biofuels provide a means to diversify fuel sources and enhance energy security. Additionally, they offer economic benefits such as job creation and reduced reliance on fuel imports. With ongoing advancements in biofuel production technology, their integration into the aviation industry is becoming increasingly feasible.

The main carotenoids of microalgae are β-carotene, lycopene, astaxanthin (Figure 4), zeaxanthin, violaxanthin, and lutein, and the most common microalgae commercially interesting for pigment production are *Dunaliella salina, Haematococcus pluvialis, Chlorella spp., Scenedesmus spp., Muriellopsis spp., Spirulina spp., Porphyridium sp.* Exceptions cannot be ignored for production of toxins with pharmaceutical applications in some species as *Ochromonas sp. and Prymnesium parvum.* produce toxins with pharmaceutical applications. The studies incorporated in this chapter are not the only contributions but several more industrial applications which may not have added significantly to the industrial output instead serve as an application. Loads of pilot studies are looking for more statistical weightage. So far, the researchers have served as a major bridge between algae and nutritional stalk to serve mankind, but a joint giant leap is awaited.

REFERENCES:

1. P. Saranraj and S. Sivasakthi. (2014). Spirulina platensis–food for future: a review. *Asian J. Pharm. Sci. Technol.*, 4(1), 26–33. https://www.researchgate.net/publication/259503619
2. Miranda, José M., Marcos Trigo, Jorge Barros-Velázquez, and Santiago P. Aubourg. 2022. "Antimicrobial Activity of Red Alga Flour (*Gelidium* sp.) and Its Effect on Quality Retention of *Scomber scombrus* during Refrigerated Storage» Foods 11, no. 7: 904. https://doi.org/10.3390/foods11070904

3. Grote B. (2019). Recent developments in aquaculture of *Palmaria palmata* (Linnaeus) (Weber & Mohr 1805): cultivation and uses. *Reviews in Aquaculture, 11*(1), 25–41. https://doi.org/10.1111/raq.12224
4. Luísa Gouveia, Anabela Raymundo, Ana Paula Batista, Isabel Sousa & José Empis (2006). Chlorella vulgaris and Haematococcus pluvialis biomass as colouring and antioxidant in food emulsions. *European Food Research and Technology, 222,* 362–367. https://doi.org/10.1007/s00217-005-0105-z
5. Yamanaka R, Akiyama, K. (1993). Cultivation and utilization of Undaria pinnatifida (wakame) as food. *Journal of Applied Phycology, 5,* 249–253. https://doi.org/10.1007/BF00004026
6. Udayan, A., Pandey, A.K., Sirohi, R. et al. (2023). Production of microalgae with high lipid content and their potential as sources of nutraceuticals. *Phytochem Rev* **22,** 833–860 https://doi.org/10.1007/s11101-021-09784-y
7. Tiziana Marino, Angela Iovine, Patrizia Casella, Maria Martino, SimeoneChianesea, Vincenzo Larocca, Dino Musmarra, Antonio Molino (2020). From Haematococcus pluvialis microalgae a powerful antioxidant for cosmetic applications. *Chem. Eng, 79,* 271–276. DOI: 10.3303/CET2079046
8. Joanna Fabrowska, Bogusława Łęska, Grzegorz Schroeder, Beata Messyasz, Marta Pikosz (2015). Biomass and extracts of algae as material for cosmetics. *Marine Algae Extracts: Processes, Products, and Applications. 16,* 681–706. https://doi.org/10.1002/9783527679577.ch38
9. Gilles Bedoux *, Kevin Hardouin * †, Anne Sophie Burlot *, Nathalie Bourgougnon (2014). Bioactive components from seaweeds: Cosmetic applications and future development. *Advances in Botanical Research, 71,* 345–378. https://doi.org/10.1016/B978-0-12-408062-1.00012-3
10. Çelebi H, Bahadır T, Şimşek İ, Tulun Ş. Use of *Dunaliella salina* in environmental applications. In Proceedings of 1st international electronic conference on biological diversity, ecology and evolution 2021 Mar 12 (p. 9411).
11. Dominguez H, Loret EP. (2019). *Ulva lactuca,* A Source of Troubles and Potential Riches. *Mar Drugs.* Jun 14;17(6):357. PMID: 31207947; PMCID: PMC6627311. DOI: 10.3390/md17060357
12. Xuhui Li, Chunxin Xia, Danni Kong, Ming Xu, Jin Zhu, Congfen He, Bingliang Wang, Junxiang Li (2022). Application of Euglena gracilis-Derived Peptides as a Cosmetic Ingredient to Prevent Allergic Skin Inflammation. *Journal of Cosmetic Science.*, 73(2). ISSN: 1525-7886.
13. Shalaby, E. (2011). Algae as promising organisms for environment and health. *Plant Signaling & Behavior, 6*(9), 1338–1350. https://doi.org/10.4161/psb.6.9.16779
14. Nuzzo D, Contardi M, Kossyvaki D, Picone P, Cristaldi L, Galizzi G, Bosco G, Scoglio S, Athanassiou A, Di Carlo M (2019). Heat-Resistant *Aphanizomenon flos-aquae* (AFA) Extract (Klamin®) as a Functional Ingredient in Food Strategy for Prevention of Oxidative Stress. Oxid Med Cell Longev. 2019 Nov 11;2019:9481390. PMID: 31827711; PMCID: PMC6885278. DOI: 10.1155/2019/9481390
15. Meinita MDN, Harwanto D, Sohn JH, Kim JS, Choi JS. (2021). *Hizikia fusiformis*: pharmacological and nutritional properties. *Foods, 10*(7), Jul 19;10(7):1660. PMID: 34359532; PMCID: PMC8306711. doi: 10.3390/foods10071660
16. Lee JH, K. B. (2019). A study on seaweed sea staghorn (Codium fragile) ethanol extract for antioxidant. *The Journal of the Convergence on Culture Technology, 5*(4), 467–472. https://doi.org/10.17703/JCCT.2019.5.4.467

17. Manisali AY. (2020). Isolation of Specific Phospholipids from *Nannochloropsis Oculata* Microalga for Cosmetic Applications. *University of South Florida*. Doctor of Philosophy (Ph.D.) "Isolation of Specific Phospholipids from Nannochloropsis oculata</" by Ahmet Yener Manisali (usf.edu)
18. Hawrot-Paw, M.; Ratomski, P.; Koniuszy, A.; Golimowski, W.; Teleszko, M.; Grygier, A (2021). Fatty Acid Profile of Microalgal Oils as a Criterion for Selection of the Best Feedstock for Biodiesel Production. *Energies, 14*, 7334. https://doi.org/10.3390/en14217334
19. Osman, M. E. H., Abo-Shady, A. M., Gheda, S. F., Desoki, S. M., & Elshobary, M. E. (2023). Unlocking the potential of microalgae cultivated on wastewater combined with salinity stress to improve biodiesel production. *Environmental Science and Pollution Research, 30*(53), 114610–114624. https://doi.org/10.1007/s11356-023-30370-6
20. Santhana Kumar V, Das Sarkar S, Das BK, Sarkar DJ, Gogoi P, Maurye P, Mitra T, Talukder AK, Ganguly S, Nag SK, Munilkumar S, Samanta S. Sustainable biodiesel production from microalgae *Graesiella emersonii* through valorization of garden wastes-based vermicompost. Sci Total Environ. 2022 Feb 10;807(Pt 3):150995. doi: 10.1016/j.scitotenv.2021.150995. Epub 2021 Oct 16. PMID: 34666095.
21. Scranton, M. A., Ostrand, J. T., Fields, F. J., & Mayfield, S. P. (2015). *Chlamydomonas* as a model for biofuels and bio-products production. *The Plant Journal, 82*(3), 523–531. https://doi.org/10.1111/tpj.12780.
22. Teresa M. Mata, António A. Martins, Nidia. S. Caetano, Microalgae for biodiesel production and other applications: A review, Renewable and Sustainable Energy Reviews, Volume 14, Issue 1, 2010, Pages 217-232, ISSN 1364-0321, https://doi.org/10.1016/j.rser.2009.07.020
23. Felix Offei, Moses Mensah, Anders Thygesen and Francis Kemausuor (2018). "Seaweed Bioethanol Production: A Process Selection Review on Hydrolysis and Fermentation". *Fermentation* 2018, *4*(4), 99; https://doi.org/10.3390/fermentation4040099
24. Song X, Kong F, Liu BF, Song Q, Ren NQ, Ren HY. Thallium-mediated NO signaling induced lipid accumulation in microalgae and its role in heavy metal bioremediation. Water Res. 2023 Jul 1; 239: 120027. doi: 10.1016/j.watres.2023.120027. Epub 2023 May 1. PMID: 37167853.
25. Chen M, Tang H, Ma H, Holland TC, Ng KY, Salley SO. Effect of nutrients on growth and lipid accumulation in the green algae *Dunaliella tertiolecta*. Bioresour Technol. 2011 Jan;102(2):1649-55. doi: 10.1016/j.biortech.2010.09.062. Epub 2010 Oct 13. PMID: 20947341.
26. Chen, Junying & Bai, Jing & Li, Hongliang & Fang, Shuqi. (2015). Prospects for Bioethanol Production from Macroalgae. Trends in Renewable Energy, 2015, Vol.1, No.3, 185-197. doi:10.17737/tre.2015.1.3.0016
27. Bouras S., Katsoulas, N., Antoniadis, D., & Karapanagiotidis, I. T. (2020). Use of Biofuel Industry Wastes as Alternative Nutrient Sources for DHA-Yielding *Schizochytrium limacinum* Production. *Applied Sciences, 10*(12), 4398. https://doi.org/10.3390/app10124398
28. Palanisamy, Karthick Murugan, Gaanty Pragas Maniam, Ahmad Ziad Sulaiman, Mohd Hasbi Ab. Rahim, Natanamurugaraj Govindan, and Yusuf Chisti. 2022. "Palm Oil Mill Effluent for Lipid Production by the Diatom *Thalassiosira pseudonana*" Fermentation 8, no. 1: 23. https://doi.org/10.3390/fermentation8010023
29. Montero MF, A. M. G. R. G. (2011). Isolation of high-lipid content strains of the marine microalga *Tetraselmis suecica* for biodiesel production by flow cytometry and single-cell

sorting. *Journal of Applied Phycology*, 23, 1053–1057. https://doi.org/10.1007/s10811-010-9623-6
30. Jackson Hwa Keen Lim, Yong Yang Gan, Hwai Chyuan Ong, Beng Fye Lau, Wei-Hsin Chen, Cheng Tung Chong, Tau Chuan Ling, Jiří Jaromír Klemeš (2021). Utilization of microalgae for bio-jet fuel production in the aviation sector: Challenges and perspective. *Renewable and Sustainable Energy Reviews.* 149, 111396. https://doi.org/10.1016/j.rser.2021.111396
31. Selvan ST, Govindasamy B, Muthusamy S, Ramamurthy D (2019). Exploration of green integrated approach for effluent treatment through mass culture and biofuel production from unicellular alga, *Acutodesmus obliquus* RDS01. *International Journal of Phytoremediation*, 21(13), 1305–1322. Epub 2019 Jun 28. PMID: 31250670. DOI: 10.1080/15226514.2019.1633255

CHAPTER 9

MICROBIAL BIOTECHNOLOGY IN FOOD INDUSTRY

Pooja Yadav and Rachna Chaturvedi

Amity Institute of Biotechnology, Amity University Uttar Pradesh, Lucknow, 226028

Abstract

Microbial biotechnology plays a pivotal role in the food industry, offering innovation solution to enhance food production, safety and quality. This field explores the diverse applications of microbial biotechnology, highlighting its impact on enhancing food quality and sustainability. Microorganisms such as bacteria, yeasts, and Molds are employed in the fermentation processes to produce a wide range of food products, including dairy, beverages, and fermented foods. These processes not only improve the nutritional and sensory attributes of foods but also extend their shelf life. Advancements in microbial biotechnology have led to the development of novel biotechnological tools, such as genetically modified microbes and enzyme technology, which are used to optimize food processing and production. The use of microbial enzymes in food processing enhances efficiency, reduces waste, and minimizes the environmental impact. Additionally, microbial biotechnology is instrumental in ensuring food safety through the development of biosensors and bio preservatives, which help in the detection and inhibition of foodborne pathogens. This chapter also addresses the challenges and future prospects of microbial biotechnology in the food industry, including regulatory considerations, consumer acceptance, and the need for sustainable practices. By harnessing the potential of microbial biotechnology, the food industry can meet the growing demand for high-quality, safe, and sustainable food products.

Keywords:
Food industry, food quality, food production, fermentation, Sustainability

INTRODUCTION

Microbial biotechnology accelerated and led to innovations in biological science, food science, sustainable agriculture, and medical science [1]. Microbial biotechnology is at the forefront of this revolution, utilizing the remarkable capabilities of microorganisms to innovate and improve various aspects of food production and processing. Microbial biotechnology harnesses microbes to transform raw materials into valuable products like food, chemicals, and biofuels through controlled fermentation processes. Microbial biotechnology enhances food processing by producing value-added products, such as enzymes, vitamins, and enzyme-rich foods, with potential for economic and health benefits [2]. Food and industrial microbiology contribute to the development of environmentally safe, sustainable, and economical alternatives for energy and food production, advancing sustainable development goals [3]. Microbial protein production from algae, yeast, fungi, and bacteria can provided a sustainable food supply with a lower environment footprint compared to plant or animal-based alternatives [4]. Microbes have been integral to food fermentation, a technique that dates back thousands of years and continues to be fundamental in producing foods such as bread, cheese, yogurt, and alcoholic beverages. Microbial enzymes, such as bacteria, yeast, and fungi, are widely used in food preparations for improving texture, and offering economic benefits to industries. Microbial enzymes and process have revolutionized food processing since the 1980s, with new enzymes and processes improving starch, sugars, proteins, fats, fibers, and flavor compounds [5]. Microbial biotechnology leverages the power of microbes to perform specific biochemical transformations, leading to the production of enzymes, bioactive compounds, and functional ingredients. Microbial production technologies in the food industry aim to produce nutritious and safe food ingredients at a low cost, while maintaining the health benefits of live microbial cultures [6].

ROLE OF MICROBES IN FOOD PRODUCTION

Microorganisms are an important part of the food industry as these are helpful in food preservation and production. Typically, fermented foods such as sourdough bread, pickles, olives, and sauerkraut, fermented meats like salami, and dairy products like yogurt and cheese are made using microbes. The food industry has recently begun using microbes extensively for the creation of chocolate, food coloring, fruit, vegetable, and meat preservation, as well as probiotics, which are beneficial to human health [7]. Fermenting microorganisms contribute to food stability, providing benefits such as antioxidants, peptide production, probiotic properties, and antimicrobial activity in fermented foods [8]. Probiotic microorganisms in non-dairy fermented foods, such as millets and cereal mixtures, may serve as affordable probiotic supplements and enhance the nutritional value of food [9]. Table 1 shows the list of microbes along with their roles in food production with examples.

Fermentation, a process driven by microbes, is a cornerstone of food production. Bacteria, yeasts, and Molds undergo metabolic transformations, converting sugars into acids, alcohols, or gases. This process not only enhances flavor, texture, and aroma but also preserves food by inhibiting the growth of spoilage organisms [10]. Moreover, microbes are employed in food safety. They act as natural preservatives, extending shelf life and reducing the risk of foodborne illnesses. Additionally, they are used in the production of food additives, enzymes, and even single-cell protein, a sustainable alternative protein source [16].

Table 1. Representing on specific roles and examples of microbes in food production

Microbes	Role in Food production	Example of food	Reference
Yeast	Fermentation of sugar to produce alcohol and CO2	Bread, beer, wine	[10]
Lactic acid bacteria	Fermentation to produced lactic acid	Yogurt, cheese, sauerkraut	[11]
Acetic acid bacteria	Oxidation of ethanol to acetic acid for vinegar production	Vinegar	[12]
Molds	Fermentation and enzyme production to enhance Flavors	Blue cheese, soy sauce	[13]
Saccharomyces cerevisiae	Primary yeast in baking and alcohol fermentation	Bread, beer, wine	[11][12]
Lactobacillus	Fermentation and probiotic function	Yogurt, kefir, pickles	[14]
Bifidobacteria	Probiotic culture for gut health	Probiotic dairy products	[15]

MICROBIAL ENZYMES IN FOOD PROCESSING:

The food processing sector greatly benefits from the use of enzymes because of their high substrate specificity, ability to function in moderate circumstances, low production of byproducts, increased yield, and reduced environmental impact. Microbial enzymes are highly beneficial in the food processing industry due to their ease of cultivation, handling, manipulation, higher efficiency, and stability, making them a promising candidate for industrial applications [17]. There are several sources of enzymes, including microorganisms, plants, and mammals. Microorganisms, especially bacteria and fungi, provide more than 50% of industrial enzymes because of their advantageous growth features, low nutritional needs, and biochemical diversity. A specialized industrial enzyme production sector in Bangladesh can benefit the environment and promote economic growth by utilizing agro-industrial waste resources [18]. Microbial enzymes, such as bacteria, yeast, and fungi, are widely used in food preparations for improving taste and texture, offering economic benefits and easy, cost-effective production. Figure 1. Represents the role of enzymes in various industries

Microbial enzymes are biological catalysts derived from microorganisms that have become indispensable tools in modern food processing. Their ability to accelerate specific chemical reactions at mild conditions offers numerous advantages over traditional chemical methods [6]. These enzymes find applications across various food industries. In baking, amylases break down complex carbohydrates, improving dough texture and bread volume. Proteases tenderize meat by hydrolyzing proteins, while lipases enhance flavor and texture in dairy products. In juice production, pectinases clarify and increase juice yield [19]. Apart from these, microbial enzymes play an important role in starch processing, producing high fructose corn syrup and glucose syrup. They are also used in the brewing industry for starch conversion and protein modification. Additionally, enzymes contribute to the production of food additives such as amino acids and vitamins. The use of microbial enzymes offers several benefits, including increased efficiency, reduced energy consumption, and improved product quality. Moreover, they are generally considered safe for consumption, making them a valuable asset in the quest for sustainable and innovative food production [20].

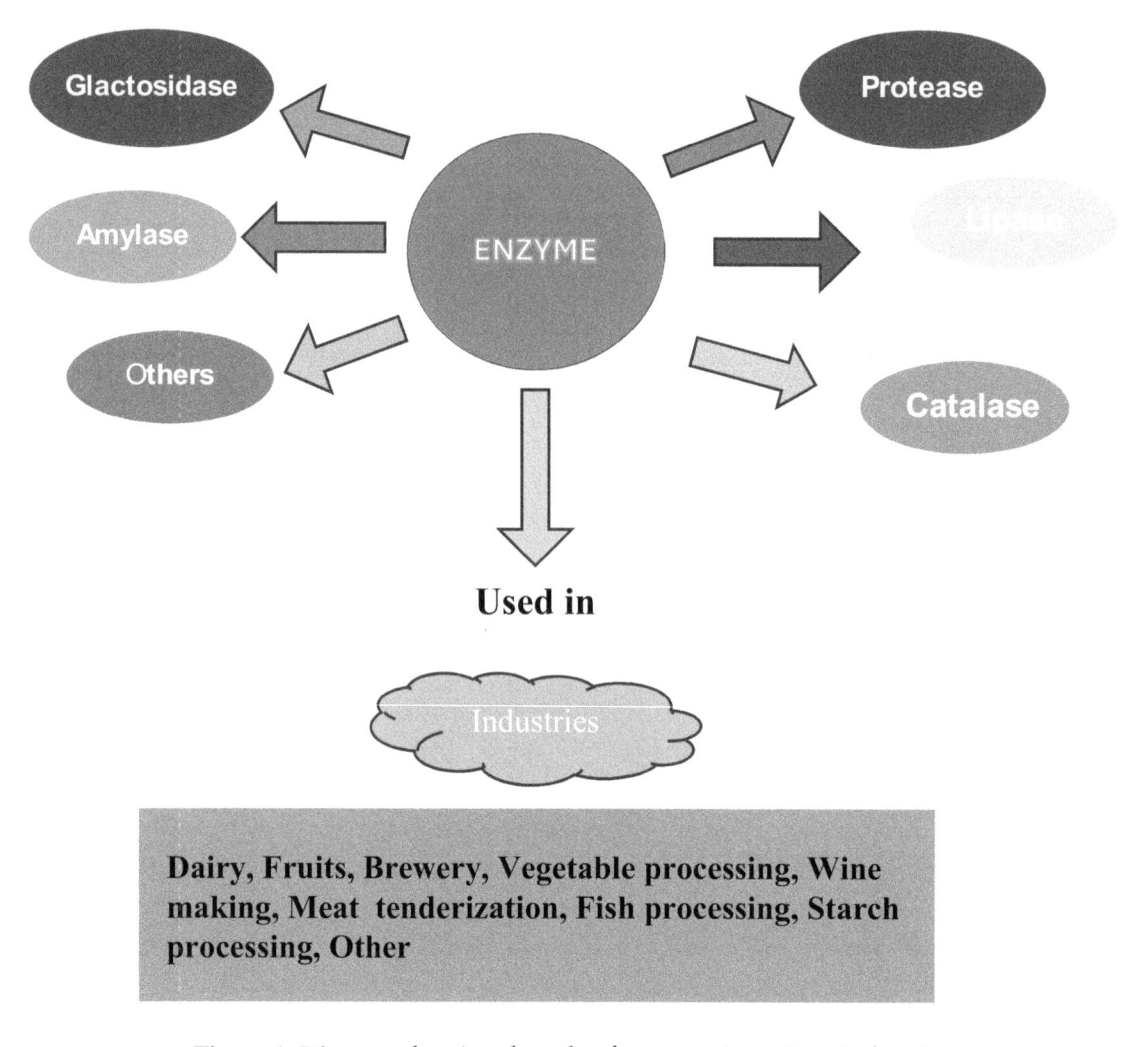

Figure 1. Diagram showing the role of enzymes in various industries

MICROBIAL SAFETY AND QUALITY CONTROL

Microbial safety and quality control is a critical aspect of ensuring the safety and integrity of various products, particularly in the food, pharmaceutical, and cosmetic industries. The potential for microorganisms to thrive in food, pharmaceutical and cosmetic products has been recognized for many years and has been the subject of debate for years [21]. Hazard Analysis and Critical Control Points (HACCP) is a systemic approach to identifying and controlling biological, chemical and physical hazards during food production. It involved identifying critical control points (CCPs) where hazards can be prevented, eliminated, or reduced to safe levels [22]. Good Manufacturing Practices (GMP) are guideline that outline the aspects of production that must be consideration to ensure food safety and quality. This includes facility cleanliness, personal hygiene, equipment maintenance, and proper handling of raw materials [22][23]. Regular microbial testing helps in monitoring and verifying the effectiveness of control measures. Common test includes total viable count, pathogen detection (e.g., Salmonella, and Listeria), and indicator organism (e.g.-coli) [24]. Sanitation standard operating procedure (SSOPs) are written procedures that outline the method for cleaning and sanitizing equipment and facilities. They are important for preventing cross-contamination and maintaining hygiene standards. Microbiological control in food processing industry is crucial for limiting microbial proliferation and stabilizing packed or canned food, ensuring microbial safety and stability. The Microbial Assessment Scheme (MAS) effectively assesses the microbial performance of Food Safety Management Systems, indicating low microorganism counts and small variations in microbial counts, contributing to food safety [25][26].

ADVANCES IN MICROBIAL BIOTECHNOLOGY

Microbial biotechnology advances in various fields, including green chemistry, food, pharmaceuticals, bioenergy, and bioremediation, are revolutionizing various industries and influencing global processes. Technological advances like microfluidics, next-generation 3D-bioprinting, and single-cell metabolomics are fueling a paradigm shift in microbiology, allowing for a full comprehension of microorganisms' chemical communication [27]. Recent advances in microbial biotechnology, such as improved expression vectors, metagenomic screening, synthetic biology, and genome editing, enable the identification and exploitation of metabolic treasures in uncultured microbes, potentially yielding tailored-made biocatalysts for multiple biotechnological applications [28]. Microbial biotechnology in food processing leverages microorganisms to transform natural substrates into diverse value-added products. This encompasses traditional fermentation processes for creating fermented foods, beverages, and dairy products, as well as advanced techniques like enzyme production, bio preservation, and functional food development. Moreover, it contributes to sustainable food systems by reducing waste, enhancing food safety, and providing innovative solutions for nutritional challenges [3]. In metabolic engineering, microbes are optimized to produce biochemicals such as biofuels and pharmaceuticals more efficiently. They act as microbial cell factories, manufacturing proteins, enzymes, and other molecules on a commercial scale [29]. The food industry benefits from innovative fermentation technologies that enhance the flavor, texture, and nutritional content of food products. Microbes also play an important role in food safety by detecting and neutralizing pathogens [30]. In the health and medicine sector, understanding the human microbiome has led to improved probiotic products and novel disease treatments. New antimicrobial agents are being discovered and engineered from microbial sources to combat antibiotic resistance. Finally, bioinformatics and data analytics are critical in mapping out microbial genomes and studying microbial community dynamics. These tools provide essential insights that drive further innovations in microbial biotechnology across all sectors [31].

CONCLUSION

Microbial biotechnology has emerged as a transformative force across various domains, particularly in the food industry. By harnessing the unique capabilities of microorganisms, it has revolutionized food production, processing, and safety, contributing to a sustainable food supply with reduced environmental impacts. The integration of microbial enzymes in food processing has enhanced efficiency, reduced energy consumption, and improved product quality, making it an invaluable asset for sustainable food systems. Advances in microbial biotechnology have facilitated the development of safe and nutritious food ingredients and innovative fermentation technologies, improving the flavor, texture, and nutritional value of food products. With the ongoing exploration of the human microbiome and the application of bioinformatics, microbial biotechnology continues to offer promising solutions to nutritional challenges, food safety, and health-related issues, ensuring a more sustainable and healthier future.

REFERENCES

1. Maurya, D., Kumar, A., Chaurasiya, U., Hussain, T., & Singh, S. (2021). Modern era of microbial biotechnology: opportunities and future prospects,17-343.
2. Subrata N. Bhowmik, Ramabhau T. Patil, (2018). Chapter 5 - Application of Microbial Biotechnology in Food Processing, Editor(s): Ram Prasad, Sarvajeet S. Gill, Narendra Tuteja, Crop Improvement Through Microbial Biotechnology, Elsevier, Pages 73-106,
3. Tripathi, C., Malhotra, J., & Kaur, J. (2022). Employing Food and Industrial Microbiology to Accelerate Sustainable Development Goals. Microsphere.
4. Matassa, S., Boon, N., Pikaar, I., & Verstraete, W. (2016). Microbial protein: future sustainable food supply route with low environmental footprint. *Microbial Biotechnology*, 9, 568 - 575.
5. Raveendran, S., Parameswaran, B., Ummalyma, S., Abraham, A., Mathew, A., Madhavan, A., Rebello, S., & Pandey, A. (2018). Applications of Microbial Enzymes in Food Industry. Food technology and biotechnology, 56 1, 16-30.
6. Demirci, A. (2005). 06 Technologies Used for Microbial Production of Food Ingredients., 158-169.
7. Mazhar, S., Yasmeen, R., Chaudhry, A., Summia, K., Ibrar, M., Amjad, S., & Ali, E. (2022). Role of Microbes in Modern Food Industry. *Vol 4 Issue 1*.
8. Sharma, R., Garg, P., Kumar, P., Bhatia, S., & Kulshrestha, S. (2020). Microbial Fermentation and Its Role in Quality Improvement of Fermented Foods. Fermentation, 6, 106.
9. Ilango, S., & Antony, U. (2021). Probiotic microorganisms from non-dairy traditional fermented foods. *Trends in Food Science & Technology*.
10. Walker, G. M., & Stewart, G. G. (2016). Saccharomyces cerevisiae in the production of fermented beverages. *FEMS Yeast Research*, 16(7), fow096.
11. Leroy, F., & De Vuyst, L. (2004). Lactic acid bacteria as functional starter cultures for the food fermentation industry. *Trends in Food Science & Technology*, 15(2), 67-78.
12. Trček, J. (2005). Microbial communities in traditional vinegar production processes. *International Journal of Food Microbiology*, 106(3), 329-345.
13. Samson, R. A., Houbraken, J., Thrane, U., Frisvad, J. C., & Andersen, B. (2010).
14. Kleerebezem, M., Boekhorst, J., van Kranenburg, R., Molenaar, D., Kuipers, O. P., Leer, R., ... & Siezen, R. J. (2003). Complete genome sequence of Lactobacillus plantarum WCFS1. *Proceedings of the National Academy of Sciences*, 100(4), 1990-1995.
15. Sánchez, B., Delgado, S., Blanco-Míguez, A., Lourenço, A., Gueimonde, M., & Margolles, A. (2017). Probiotics, gut microbiota, and their influence on host health and disease. *Molecular Nutrition & Food Research*, 61(1), 1600240.
16. eshome, E., Forsido, S. F., Rupasinghe, H. P. V., & Olika Keyata, E. (2022). Potentials of Natural Preservatives to Enhance Food Safety and Shelf Life: A Review. *TheScientificWorldJournal*, 2022,9901018.
17. Mehta, P.K. and Sehgal, S., 2019. Microbial enzymes in food processing. *Biocatalysis: Enzymatic Basics and Applications*, pp.255-275.
18. Hossain, I., Mitu, I., Hasan, M., & Saha, S. (2023). Industrial enzyme production in Bangladesh: current landscape, scope, and challenges. *Asian Journal of Medical and Biological Research*.
19. Shinde, Dr. & Deshmukh, Shubham & Bhoyar, Mahesh. (2015). Applications of Major Enzymes in Food Industry. Indian Farmer. 2. 498-502.
20. Gurung, N., Ray, S., Bose, S., & Rai, V. (2013). A broader view: microbial enzymes and their relevance in industries, medicine, and beyond. *BioMed research international*, 2013, 329121.

21. Nemati, M., Hamidi, A., Maleki Dizaj, S., Javaherzadeh, V., & Lotfipour, F. (2016). An Overview on Novel Microbial Determination Methods in Pharmaceutical and Food Quality Control. *Advanced pharmaceutical bulletin*, 6(3), 301–308.
22. Mortimore, S., & Wallace, C. (2013). *HACCP: A practical approach*. Springer Science & Business Media.
23. World Health Organization. (2007). Quality assurance of pharmaceuticals: a compendium of guidelines and related materials: vol. 2: Good manufacturing practices and inspection, 2nd ed. World Health Organization.
24. American Public Health Association. (2015). Compendium of Methods for the Microbiological Examination of Foods (5th ed.).
25. Marriott, N.G. & Schilling, Wes & Gravani, R. (2017). Principles of Food Sanitation. 10.1007/978-3-319-67166-6.
26. Jacxsens, L., Kussaga, J., Kussaga, J., Luning, P., Spiegel, M., Devlieghere, F., & Uyttendaele, M. (2009). A Microbial Assessment Scheme to measure microbial performance of Food Safety Management Systems. *International journal of food microbiology*, 134 1-2, 113-25.
27. Nai, C., & Meyer, V. (2017). From Axenic to Mixed Cultures: Technological Advances Accelerating a Paradigm Shift in Microbiology. *Trends in microbiology*, 26 6, 538-554.
28. Santero, E., Floriano, B., & Govantes, F. (2016). Harnessing the power of microbial metabolism. *Current opinion in microbiology*, 31, 63-69.
29. Quinn Zhu, Ethel N Jackson, (2015) Metabolic engineering of Yarrowia lipolytica for industrial applications, Current Opinion in Biotechnology, Volume 36, Pages 65-72.
30. Voidarou, C., Antoniadou, M., Rozos, G., Tzora, A., Skoufos, I., Varzakas, T., Lagiou, A., & Bezirtzoglou, E. (2020). Fermentative Foods: Microbiology, Biochemistry, Potential Human Health Benefits and Public Health Issues. *Foods (Basel, Switzerland)*, 10(1), 69.
31. Patangia, D. V., Anthony Ryan, C., Dempsey, E., Paul Ross, R., & Stanton, C. (2022). Impact of antibiotics on the human microbiome and consequences for host health. *MicrobiologyOpen*, 11(1), e1260.

CHAPTER 10

GOOD BACTERIA AND MICROBIOME FOR HUMAN HEALTH

Aviral Singh, Jyoti Prakash, Ruchi Yadav and Rachna Chaturvedi

Amity Institute of Biotechnology, Amity University Uttar Pradesh, Lucknow, 226028

Abstract

The mortal body has around unimaginable human cells. However, our microbiome is estimated to be more than human cell microbial cells which consist of bacteria, viruses, and other microorganisms that colonize the human body coexisting peacefully with their host. Bacteria are tiny single-celled organisms. They are known to be found in almost every single place on earth and are essential for the earth's well-being. The human body consists of bacteria and microbes and according to studies most of them are harmless and some are even helpful for the survival of human beings. They play an essential role in day-to-day bodily function which promotes sustainable good health Microbiomes play a crucial role in our health by helping digestion and benefiting our immune system and wide aspects of health in the human body. This paper mainly focuses on good bacteria and microbiomes which are essential and promote human health.

Keywords:
Microbiome, Bacteria, Human Beings, Peacefully, Mortal

INTRODUCTION

Microbiomes are the collection of all microbes that are both beneficial and potentially harmful for human health, such as bacteria, fungi, viruses, and their genes, that live on and in our bodies naturally but as we know microbes are so small in size that they cannot be seen with naked eyes it requires the help of a microscope to see them but as small as they were they contribute way to big in human health and wellness. The microbiota study has been moved forward rapidly in the last many years and has now become a great study for scientific research human microbiome research has emerged from the environmental microbiome study it is kind of evident that the human body is also an ecosystem where trillions of microscopic organisms coexist peacefully with their host.

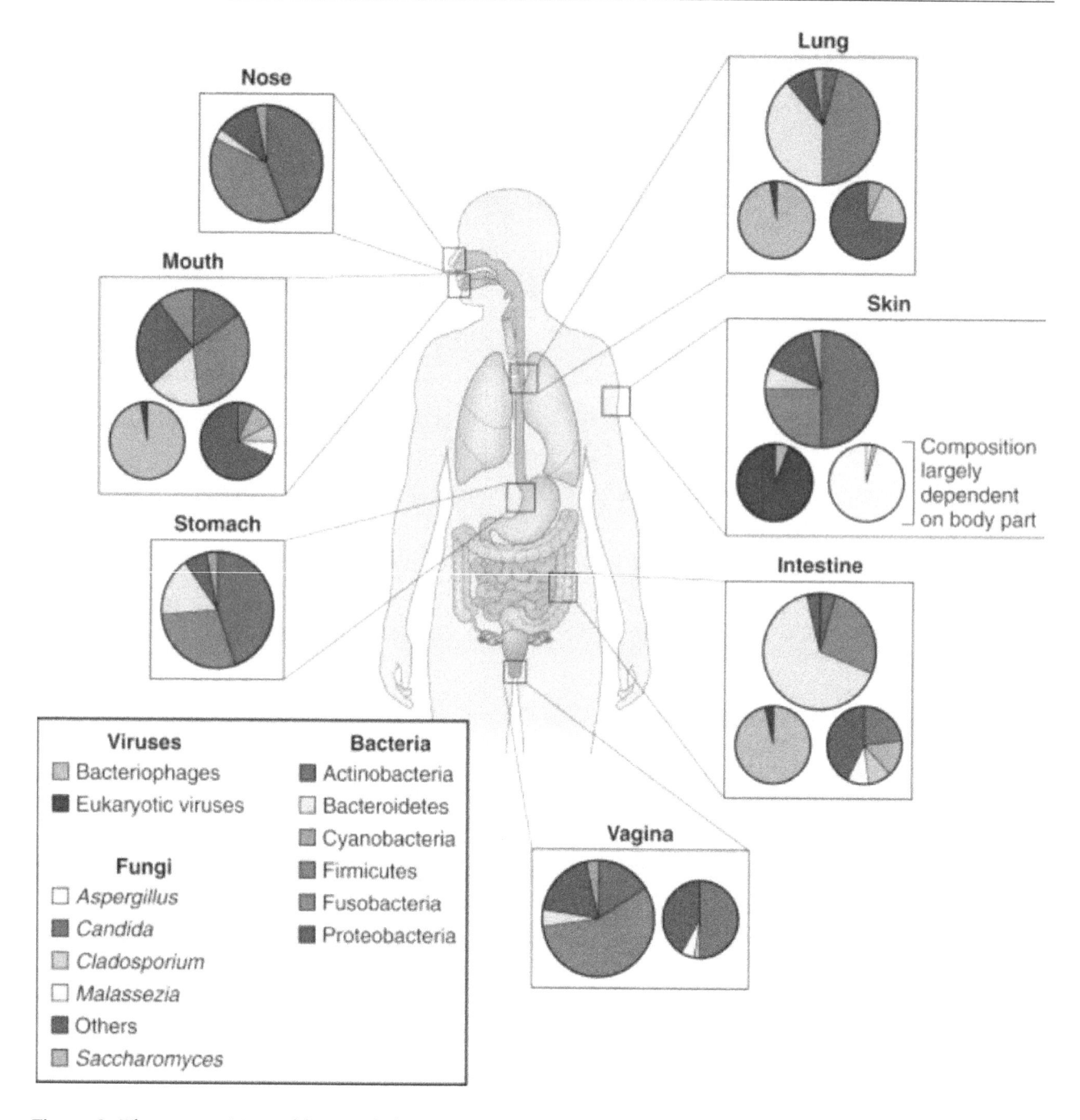

Figure 9: The composition of bacterial, fungal, and viral microbiota at distinct body sites. This figure shows the distribution and relative abundance of bacterial, fungal, and viral communities at different sites on the human body that are exposed to the external environment. Bacterial composition is represented by the six most abundant phyla, fungal composition by the most prominent genera, and viral composition as bacteriophages or eukaryotic viruses. (Marsland BJ, Gollwitzer ES. Host-microorganism interactions in lung diseases. Nat Rev Immunol. 2014; 14:827–35.) (3)

The microbes population is about 10 times larger than the total number of somatic and germ cells in our body. They protect our body against pathogens and help us to digest our food effectively and efficiently which as a result produces energy and it also helps our immune system to develop. There is a vast no. of microbes and bacteria found in the human body which promote and affect the hosts' health (1). Microbiota are known to be found all over the human body, starting from the skin to the gut, and in the past blood is considered a sterile environment from the start of clinical microbiology, it has come to our knowledge that microbial families colonize particular places in the human body, for example:- staphylococci resides the skin, Escherichia coli resides the colon and lactobacilli reside in the vagina (1). The most colonized site is the gastrointestinal tract (GIT). >70% of all the microorganisms in the human body reside in the colon. The gut microorganisms make essential micronutrients, vitamins, and enzymes, which allow us to digest foods and absorb various important nutrients (2). The disease-causing microbes have an apparent effect which makes them identified earlier than beneficial microbes. The essential microbes protect the human body against the colonization of harmful microorganisms and serve as a barrier to reduce exposure to harmful agents. In this chapter, we highlight the microbiome and good bacteria found in different niches of the human body with a detailed explanation of how they are beneficial or harmful to human health.

THE HUMAN GASTROINTESTINAL TRACT MICROBIOME

The human microbiome is the main focus of one of the most important research areas of our time, and most of the effort is focused on the gastrointestinal tract, which contains almost all the microbes we encounter (4). (Human Body) as the gastrointestinal tract run from the oral cavity to the anus and the microorganism which populates the human gastrointestinal tract has a essential role in managing human health and diseases. The oral cavity is the main entry point into the human body and therefore the microbes living in the oral cavity can spread to different parts of the body and cause diseases, therefore the composition of the oral cavity plays a decisive role in ensuring immunity to human health, for example, during nitrate metabolism of the microbe, nitrates are. Reduced to nitrites and then nitrite is transformed. to nitric oxide, which has antimicrobial properties and is important for vascular health (5). The complex symbiotic relationship between the gut microbiome and its host is strongly influenced by diet and nutrition and, when optimized, can be extremely beneficial for digestion, nutrient absorption, and immune system health (6). Among all body sites, the highest taxonomic unit and genetic content are observed in fecal samples representing the intestinal tract, and most of the microbes colonizing here are anaerobes, with a content of approx. 4000 bacterial species with an estimated cell count of approximately 1012 cells per gram (1). The maintenance of homeostasis in the human body is also maintained by the bacteria inhabiting the intestinal tract an imbalance in their compositional state can produce disease state.The phyla Actinobacteria, Firmicutes, Proteobacteria, Bacteroidetes, Cyanobacteria, and Fusobacteria dominate the human body. The relative abundance of these lineages varies in different body parts and varies between individuals depending on their diet, geographic location, and age. Although Firmicutes and Bacteroidetes are the two main bacterial species that account for up to 90% of the gut microbiome, they dominate the normal gut microbiome and are able to efficiently digest complex dietary polysaccharides (6). Almost 200 different genera can be found in the genus Firmicutes, some of which include Bacillus, Lactobacillus, Enterococcus, and Clostridium. In addition, species such as lactobacilli are beneficial for health, on the other hand, some of them, such as Staphylococcus aureus and Clostridium perfringens, are harmful to the body (6). to grow Meanwhile, the less common Actinobacteria family is mostly represented by Bifidobacterium, which is known to be positive for health. However, the composition of the gut microbiota can vary between healthy individuals, and the Indian and British microbiomes may differ.Intestinal bacteria such as Lactobacillus, Enterococcus, and Bifidobacterium are important for maintaining epithelial integrity, strengthening the intestinal barrier, and protecting against chemical-induced epithelial barrier breakdown (7). They also play an important role in regulating digestion in the gastrointestinal tract. Commensal bacteria play a key role in the processing of nutrients and many types of metabolites such as amino acids, bile acids, short-chain fatty acids, and many others. Includes energy storage and metabolic efficiency. Some of these members maintain the integrity of the intestinal epithelium, which ultimately functions as an important immune system against pathogenic bacteria and prevents bacterial invasion. Numerous Gram-positive facultative anaerobic or microaerophilic Lactobacillus and other Bifidobacterium (actinobacteria) species found in the gastrointestinal tract, such as Lactobacillus brevis and Bifidobacterium dentium, can metabolize glutamate to produce gamma-aminobutyric acid, a major inhibitor of GABA, human central nervous system. Although the gastrointestinal tract microbiome is beneficial for human health future studies will likely shed some more light on the topic GIT microbiome which leads to further advancement in health (7).

THE HUMAN RESPIRATORY TRACT MICROBIOME

Due to the recent period because of the COVID-19 pandemic efforts were made to understand the microbiome of the respiratory tract in a better and efficient way from birth to adulthood of how they interact with environmental external pathogens and the host immune system. Our body is exposed to millions of different kinds of microorganisms in our daily routine like breathing. The microbiome mainly acts as a gatekeeper for the pathogens by providing resistance to these they were also involved in the maturation and maintenance of homeostasis of respiratory physiology and immunity. The human respiratory tract harbors many microbes not as much as the gastrointestinal tract which we have mentioned before the respiratory tract starting from the nostrils to the alveoli in our lungs they were thought to be beneficial to human health as it primer the immune system which overall strengthens the immunity and protect our body with harmful pathogen attack from the environment as they disrupt the respiratory tract environment as we inhale in the air while breathing which carries all types of elements and microbes into our body through nostrils affecting our upper respiratory tract (nose; nasal cavity; pharynx; larynx) and lower respiratory tract (trachea; bronchi; bronchi and alveoli) (8). The upper respiratory tract is an interconnected system of cavities consisting of the nostrils, pharynx, and oropharynx, which are connected to the cavity of the larynx and the middle ear through the Eustachian tube (8). these areas were surrounded by mucous which was harbored by a wide range of bacteria like Firmicutes, Bacteroidetes, Proteobacteria, Actinobacteria, and Fusobacteria (9). The environment created by the upper respiratory system is directly linked with the microbiome as it influences the microenvironment of the microorganism colonizing in the respiratory tract as the function of the upper respiratory tract is to filter, warm, and moisten the air before it reaches the lungs (1). The composition of the microbiome present in different regions is directly influenced by the microenvironment created by the upper respiratory system. The pathogen competes with microbes for attachment sites, and nutrients and makes it difficult to multiply microbes and cause disease but the normal microbiome prevents the formation of the pathogen in the respiratory tract by competing with the harmful pathogens and ultimately prevents the pathogenic microorganism formation. The lower respiratory tract basically consists of the trachea and the parts present inside the lungs the microbiome present in the lungs is determined by the conditions present in the early year of life which changes with age, diet, surrounding environment, and also the use of antibiotics or any type of antimicrobial medicine taken from the birth as lungs do not have similar microbial condition throughout therefore they were depending on many factors like (A) Microbial Inhalation; (B) Microbial Exhalation (C) and Local growth condition and if there is a low microbial exhalation/elimination it will affect the regional growth conditions and creates dysbiosis which ultimately leads to high-risk lung disease (10). The lower respiratory tract infections are relatively few but have a high mortality rate. It has been noticed that changes that alter or affect the symbiosis between the bacteria and the host lead to the condition known as dysbiosis which further leads to diseases like asthma and all kind of respiratory infections and allergies as people have developed a relationship with their symbiotic bacteria which is beneficial as well as good for the human health (10).

THE HUMAN REPRODUCTIVE TRACT MICROBIOME

The human reproductive tract microbiome is not as well researched as the gastrointestinal tract(gut) microbiome, as the gut microbiome provides many beneficial factors to the human body by digesting food and strengthening immunity similarly it has been believed that commensal colonization in reproductive tract microbiome also provide a pathogenic free and strengthen healthy immune state to reproductive tract which is beneficial to many extends for human health (11). It has been agreed for a long period of time that the human vagina is colonized by the commensal bacteria predominantly Lactobacilli by creating an acidic environment by the Lactobacilli spp. which act as the very first barrier in protecting the upper reproductive tract against the harmful pathogenic microbes in females and lactobacilli as a member of the genus Firmicutes, which typically dominate these microbiomes, can act as probiotics and prevent the overgrowth of other bacterial species (11). Vaginal microbes are certainly not constantly changing due to estrogenic levels, puberty, menstruation, differences in sexual activity, and clinical symptoms of bacterial vaginosis, when estrogen levels and vaginal pH change during pregnancy, therefore the pH shows signs. The vaginal microbiota becomes less diverse and richer throughout pregnancy, shifting toward a population dominated by Lactobacillus (12) (13). Aagaard *et al.* (2012) (13)showed that while the microbiome of women who were getting closer to term recovered to its non-pregnant form, the variety and richness of the microbiome decreased during pregnancy in the 24 pregnant women's vaginal samples compared to the 60 non-pregnant controls. (12)increase the presence of lactobacilli as the gestation period increases (11) (1).

It has been long thought that the uterus was sterile which means no microbes reside or colonize there which has been changed by the finding from the Human Microbiome Project (HMP), The presence of an active microbiome in the female reproductive system has long been recognized. The vaginal environment has received the most attention, but evidence showing the rest of the female reproductive axis is not sterile has been mounting for decades. In fact, almost all of the more than 20 completed investigations have discovered that the uterine cavity contains a tiny but active microbiome. It is noteworthy that a significant number of these studies obtained their samples during surgery by using trans-fundal collection techniques, which eliminates the possibility of contamination from passing through the endocervical canal or vagina. In the vast majority of these investigations, any bacteria found were identified through the application of conventional culture methods. Which is found and states that the uterine cavity is colonized by its unique microbiome (14). The human Uterine cavity is colonized by abundant Lactobacilli, Bacteroides, Gardnerella, and Prevotella and also has Firmicutes and Actinobacteria in the vaginal microbiome (11) (15). There has been no relation found between the vaginal and uterine microbiome and the function of the uterine microbiome is still unknown (15).

THE HUMAN SKIN MICROBIOME

The skin is the largest organ of the body that comes into contact with the external environment and surprisingly the most interaction with the environment skin microbiome remains stable in harsh conditions. That is why it is known as the first line of defense against pathogens (16). It functions as both As skin is home to millions of bacteria, fungi, and viruses they are the essential part of our body as the gut microbiome is an essential part of our body which digests our food and strengthens our immune system. The dominant resident species of skin bacteria are commensals. Together with immune cells and keratinized skin cells (replaced every four weeks), these are responsible for the appropriate skin immune barrier functioning (17). the skin microbiome is also responsible for protecting against pathogens, strengthening and making our immune system aware of external threats, and breaking down natural products as mentioned above the skin microbiome protects the human body from pathogenic invasion from the environment that it acts as a barrier and in any case if this barrier is dissolved or been broken or when the commensalism between the microbes and the skin is disturbed it causes skin diseases (18). About (>90%) of bacteria in the human skin microbiome are classified into four types: Actinobacteria (52%), Firmicutes (24%), Proteobacteria (16%), and Bacteroidetes (6%) (19). According to estimates, representatives of *Cutibacterium*, *Staphylococcus*, and *Corynebacterium* genera, isolated from almost all skin areas, may constitute 45 to 80% of the entire skin microbiome (20) (19). fungi—mainly belonging to the Ascomycota and Basidiomycota types—also form part of the skin microbiome. The dominating genus is *Malassezia*. The highest level of fungi diversity has been observed on the feet, colonized by *Aspergillus*, *Cryptococcus*, *Rhodotoula*, and *Epicoccum*. However, bacteria are still the most dominant group of the skin microbiome (19) (21)

CONCLUSION

In conclusion, Microbiome present in our body whether they are Gastro intestinal tract microbiome which has almost 90% of the microbes of the whole human body which helps us in digesting our food which in end result produces energy for our daily bodily functional and strengthen our immunity or if they are Respiratory tract microbiome which has come to lot of attention since COVID-19 pandemic which make the researcher to go more towards microbes resides in our respiratory tract or if they are reproductive tract or skin microbiome which play an crucial role in maintaining the healthy environment for our reproductive tract which contributed toward pregnancy and healthy vaginal site they also get affected by sudden changes in pH levels which cause diseases and the microbiome that protects out body from harmful pathogenic microbes environment entering our body through small pores in our skin where microbiome act as barrier respectively they all play a crucial role in maintaining our health they were present in different niches of our body protecting it against pathogenic microbes and maintaining the healthy environment inside.

Microbiome research which was been conducted and being under process also states the importance of human microbes' interaction to be successful which promotes human health and various disease-causing processes which points towards a whole new direction that the microbiome is the potential target for disease management.

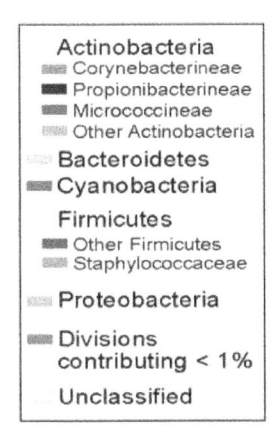

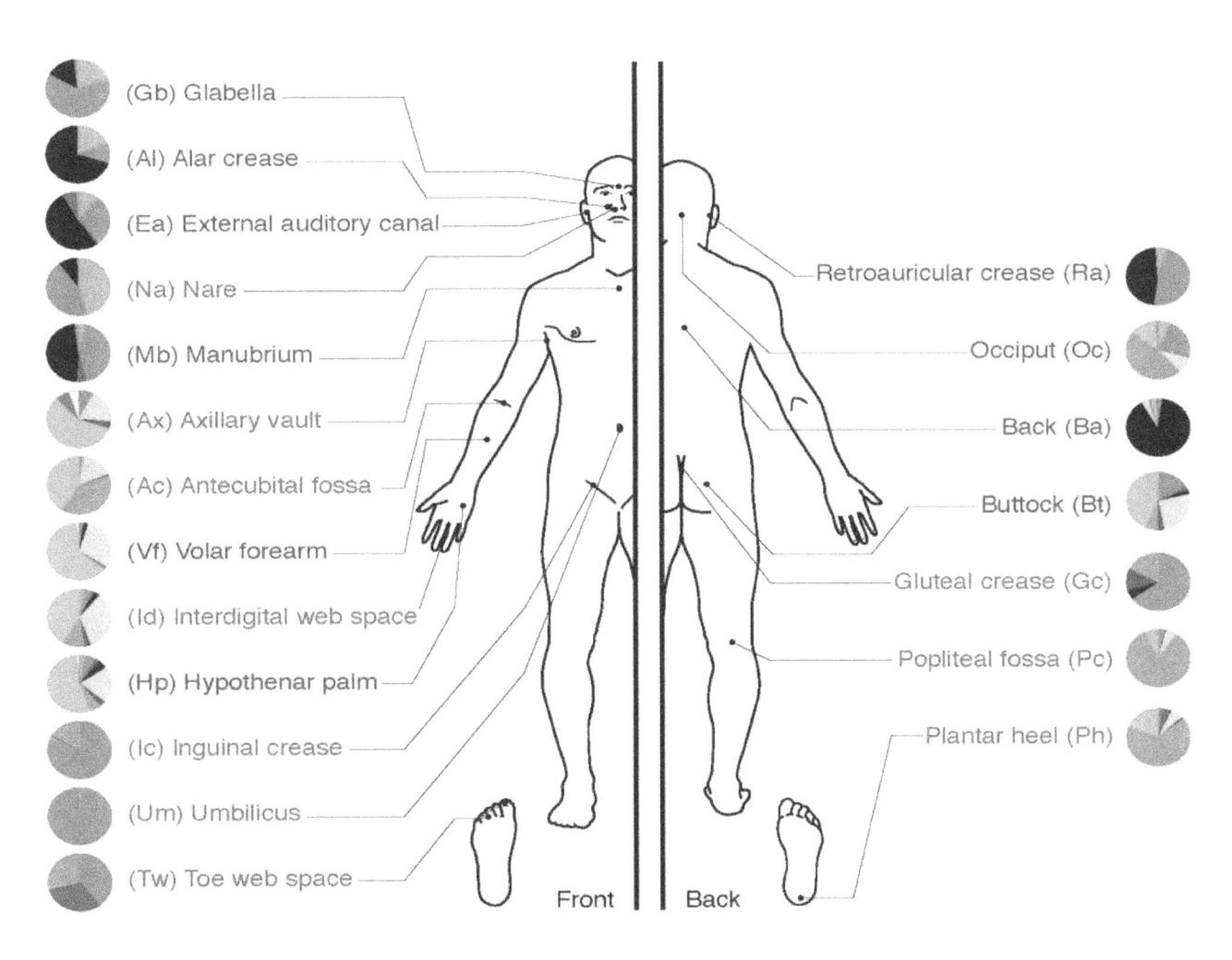

Good Bacteria and Microbiome for Human Health 107

Figure 10: Types of bacteria found on human skin (22)

REFERENCES

1. Methé BA, Nelson KE, Pop M, Creasy HH, Giglio MG, Huttenhower C, Gevers D, Petrosino JF, Abubucker S, Badger JH, Chinwalla AT. A framework for human microbiome research. Nature. Jun 14, 2012, p. 486.
2. Microorganisms with claimed probiotic properties: an overview of recent literature. S, Fijan. 2014, International Journal of Environmental Research and Public Health., pp. 4745-67.
3. Host-microorganism interactions in lung diseases. (Marsland BJ, Gollwitzer ES. 2014, Nat Rev Immunol., Vol. 14, pp. 827–35.
4. Microbiome and human health: Current understanding, engineering, and enabling technologies. Aggarwal N, Kitano S, Puah GR, Kittelmann S, Hwang IY, Chang MW. 2022, Chemical Reviews.
5. The oral microbiome in health and disease. WG., Wade. 2013, Pharmacological research., pp. 137-43.
6. Rinninella E, Raoul P, Cintoni M, Franceschi F, Miggiano GA, Gasbarrini A, Mele MC. What is the healthy gut microbiota composition? A changing ecosystem across age, environment, diet, and diseases. Microorganisms. Jan 10, 2019, p. 14.
7. Zoetendal EG, de Vos WM. Effect of diet on the intestinal microbiota and its activity. Current opinion in gastroenterology. Mar 1, 2014, pp. 189-95.
8. Man WH, de SteenhuijsenPiters WA, Bogaert D. The microbiota of the respiratory tract: gatekeeper to respiratory health. Nat Rev Microbiol. May 15, 2017, pp. 259-270.
9. Santacroce L, Charitos IA, Ballini A, Inchingolo F, Luperto P, De Nitto E, Topi S. The Human Respiratory System and its Microbiome at a Glimpse. Biology (Basel). Oct 1, 2020, p. 9.
10. Dickson RP, Erb-Downward JR, Freeman CM, McCloskey L, Beck JM, Huffnagle GB, Curtis JL. Spatial variation in the healthy human lung microbiome and the adapted island model of lung biogeography. Annals of the American Thoracic Society. Jun 12, 2015, pp. 821-30.
11. Prince AL, Chu DM, Seferovic MD, Antony KM, Ma J, Aagaard KM. The perinatal microbiome and pregnancy: moving beyond the vaginal microbiome. Cold Spring Harbor perspectives in medicine. Jun 1, 2015, p. 5.
12. Identification and evaluation of the microbiome in the female and male reproductive tracts. Rivka Koedooder, Shari Mackens, Andries Budding, Damiat Fares, Christophe Blockeel, Joop Laven, Sam Schoenmakers. 3, 2019, Human Reproduction Update, Vol. 25, pp. 298–325.
13. Aagaard K, Riehle K, Ma J, Segata N, Mistretta T-A, Coarfa C, Raza S, Rosenbaum S, Van den Veyver I, Milosavljevic A. A metagenomic approach to characterization of the vaginal microbiome signature in pregnancy. 2012.
14. "Reproductive tract microbiome in assisted reproductive technologies.". Franasiak, Jason M., and Richard T. Scott Jr. 6, 2015, Fertility and sterility, Vol. 104, pp. 1364-1371.
15. Baker JM, Chase DM, Herbst-Kralovetz MM. Uterine microbiota: residents, tourists, or invaders? Frontiers in immunology. Mar 2, 2018, p. 208.
16. "Temporal stability of the human skin microbiome.". Oh, Julia, Allyson L. Byrd, Morgan Park, Heidi H. Kong, and Julia A. Segre. 4, 2016, Cell, Vol. 165, pp. 854-866.
17. "Microbial ecology of the human skin.". Cundell, Anthony M. 1, 2018, Microbial ecology, Vol. 76, pp. 113-120.
18. Belkaid Y, Segre JA. Dialogue between skin microbiota and immunity. Nov 21, 2014, pp. 954-9.

19. "Human Skin Microbiome: Impact of Intrinsic and Extrinsic Factors on Skin Microbiota". Skowron, Krzysztof, Justyna Bauza-Kaszewska, Zuzanna Kraszewska, Natalia Wiktorczyk-Kapischke, Katarzyna Grudlewska-Buda, Joanna Kwiecińska-Piróg, Ewa Wałecka-Zacharska, Laura Radtke, and Eugenia Gospodarek-Komkowska. 3, 2021, Microorganisms, Vol. 9, p. 543.
20. Samaras, Samantha, and Michael Hoptroff. The Microbiome of Healthy Skin.". Skin Microbiome Handbook: From Basic Research to Product Development. 2020, pp. 1-32.
21. Buerger, Sandra. "The Skin and Oral Microbiome: An Examination of Overlap and Potential Interactions between Microbiome Communities.". Skin Microbiome Handbook: From Basic Research to Product Development. 2020, pp. 45-57.
22. Ahern, Holly. The Human Skin Microbiome Project. Microbiology: A Laboratory Experience. s.l.: Milne Open Textbooks, 2018.

CHAPTER 11

INTEGRATING BIOINFORMATICS INTO BIOREMEDIATION STRATEGIES

Palak Sachdeva, Ruchi Yadav, Rachna Chatuvedi and Jyoti Prakash

Amity Institute of Biotechnology, Amity University Uttar Pradesh, Lucknow, 226028

Abstract

Bioinformatics means the fusion of bioscience and computer science that is based on the use of computational approaches to analyze, govern, and store biological data. It also sheds light on fundamental and microscopic levels to be applied in modern biotechnology. For the analysis and interpretation of biological data, several tools and software are developed. Bioremediation, the latest technology researches microbial capacity for Biodegradation of xenobiotic mixtures. Microbial cells exhibit a great variety of contamination deterioration potentials that effectively reimpose natural environmental conditions. Some areas of proteomics and genomics are used in the study of Bioremediation. Bioinformatics needs the study of the genome of microorganisms, proteomics, computational bioscience, and phylogeny to know the structure of databases that are used for better upliftment. This paper points to the use of Bioinformatics ideas tested in the area of Bioremediation.

Keywords:
bioinformatics, genomics, proteomics, microbiology, bioremediation

INTRODUCTION

In today's times, environmental pollutants are a crucial universal concern, given their unpleasant effects and hazardous xenobiotic compounds. [1] [2] There are several Polynuclear Aromatic Hydrocarbons (PAHs), xenobiotic compounds, [3] nitro-aromatic compounds and chlorinated compounds which proved to be extremely cancerogenic, mutagenic, and poisonous to life forms. [4] Due to its high distribution, many microorganisms are considered the main source of the improvement of most environmental pollutants in the biogeochemical cycles of all living organisms. These organisms exhibit a wide range of pollution-reducing abilities and can restore the natural environment. [5] Substantial approaches in molecular technologies enable universal profiling of gene expression; whole-genome study of DNA (popularly known as genomics), RNA definition (known as transcriptomics), and protein study (called proteomics) create a chance for thorough physiological manifestations of the organism. From a Bioinformatics point of view, Bioremediation offers many interesting possibilities. This field needs the aggregate of massive amounts of statistics out of different resources; arrangement and chemical reactions of natural substances; structures and capabilities of enzymes (proteins); environmental microbiology; and genomics. The collection of massive amounts of records on genes and enzymes(proteins) recognizes the field of bioscience from a 'systems' perspective. In this regard, biological systems are said to be made up of elements in a compound manner, features of which can be studied separately and later on summing up a system as a complete unit. Similarly, the records associated with Bioremediation are collected in public records. This permits the study of bioremediation from a systems point of view that adds a traditional approach that focuses on separate components

BIOINFORMATICS

Bioinformatics has surfaced into a complete multi-disciplinary subject that combines facts from information and computer technology and is related to natural sciences. Bioinformatics is a program for data design, information handling, information storage, data retrieval, and worldwide connections. Bioinformatics means the documentation, analysis collection, examination, and investigation of gene and protein data. It includes databases of sequences and structural information. Bioinformatics aims at maintaining databases regularly thereby submitting new entries. Biodiversity, cell metabolism, proteome and genome analysis, vaccine and medicine development are some of the sectors of which bioinformatics is an integral element.

Organization process in bioinformatics includes the creation and conservation of databases of biological information.

Analyzing processes in bioinformatics includes the development of styles to prognosticate nucleic acid sequences, the construction of protein models, and The development of ancestral trees for analyzing evolutionary connections.

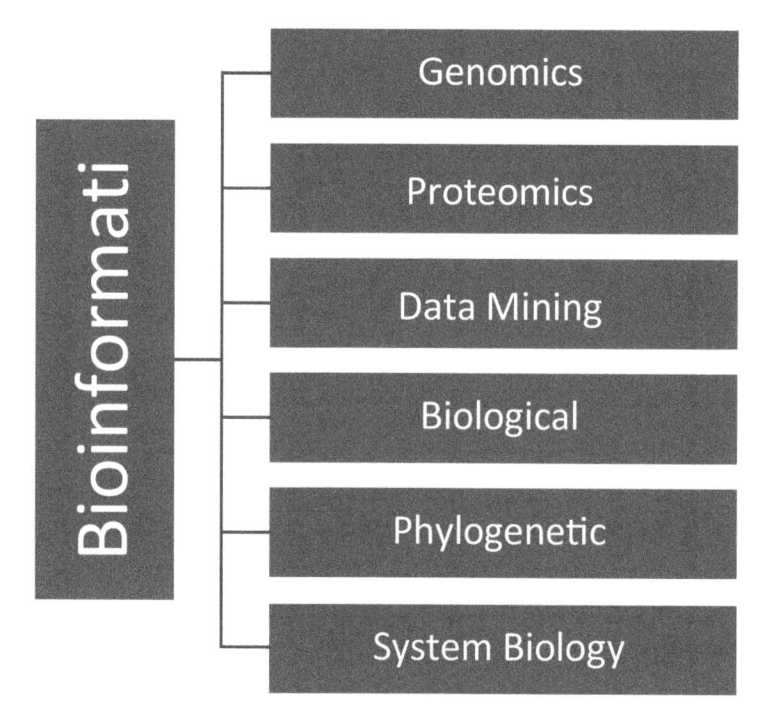

Figure 1

BIOREMEDIATION

Bioremediation is a process that uses metabolic techniques to transform pollutants so that they are not in an undesirable state. In some cases, pollutants are an important part of the life processes that aim at providing energy and carbon to microorganisms. Whereas, in some cases, it is first converted to an intermediate which serves as an important carbon and energy source. Lately, the bioremediation technique aims to immobilize inorganic pollutants such as heavy metals. [6]

Integral Bioremediation is performed unaltered focusing on the living techniques performed via native organisms. Elemental Bioremediation is a technique that is most relevant in naturally contaminated soil containing PAH (Polyhydroxy aromatic hydrocarbons). [7] Bio-stimulation helps in increasing the action of organically transpiring microbial cells that are present for Bioremediation.

Bioremediation is based on the existence of suitable microbial cells in proper quantities and combinations with natural conditions. Bioremediation aims at converting or degrading contaminants to less harmful chemicals. [8]

Off-site Bioremediation is the technique used for the analysis of ground in which soils are removed or debris is taken out from the subsurface and then disinfected. [9] Genetically modified microorganisms help in bioremediation. Bioremediation provides many interesting prospects from a bioinformatics perspective where it needs a huge extent of information on natural and man-made compounds. The foundation of microarrays, genomic analysis, and microorganism destruction pathways of compound aggregate are provided by data such as functional genomics, proteomics, evolutionary biology, etc. [10]

PROTEOMICS IN BIOREMEDIATION

In 1995, the terms 'proteome' and 'proteomics' were discovered, which is an important post-genomic property that surfaced since the evolution of massive and complicated databases of genome sequences. The proteomic investigation is mainly important as the noticed trait is an immediate outcome of the activity of the polypeptide instead of the gene order. Conventionally, this innovation relies on extremely systematic techniques of partition adopting 2-dimensional Polyacrylamide Gel Electrophoresis (2D-PAGE) and present tool of Bioinformatics in co-occurrence with mass spectrometry. Although, 2D-PAGE has been studied to be a restricted pathway for fundamental and hydrophobic membrane proteins in compartmental proteomics. The protein structure of the extrinsic proteins is of huge concern, especially in PAH degradation in the field of bioremediation. [11] The advancements in 2D-PAGE for utilization in constituent proteomics have been constructed by launching different access for multi-dimensional polypeptide description innovation. [12]

PROTEOMICS TOOLS AND TECHNIQUES APPLIED IN BIOREMEDIATION

Depending on the environment, a microorganism's biological protein composition changes.[13]

Due to organisms' adaptive reactions, the presence of hazardous substances in the environment may modify how physiologically they react to external stimuli. The development of proteomics has made it possible to thoroughly investigate regional variations in protein composition and identify key proteins linked to microbial responses in anatomical settings. Proteomics has been used to examine the biology of Bacillus subtilis and its effectiveness has been evaluated. A large-scale proteomic study has been done to understand the genesis of biofilms, the distribution of stimulus and regulons, the resistance of bacilli to severe conditions, and comprehensive proteome mapping

Because of their adaptability and simplicity in culture, the bacterial genus Pseudomonas and related genera have been extensively explored, making them particularly dynamic in pollutant degradation and aerobic decomposition. [14] Planktonic or sessile, heterotrophic, autotrophic, or anaerobic Pseudomonas can all exist together. Many of them can both break down aliphatic and aromatic hydrocarbons as well as withstand non-degradable hazardous contaminants. Since pseudomonas have a low requirement for nutrients, a variable metabolic rate, and the capacity to build biofilms, they are mostly researched from a proteomics perspective. A gram-negative bacterium called *Shewanella oneidensis* dwells in anoxic conditions. Its ability to reduce a variety of organic chemicals, metal ions, etc., is directly related to its bioremediation capacity. Proteomics studies have been conducted to learn more about bacteria in general, progress the definition of databases, and offer a variety of chromatographic and mass spectrometric proteomics methodologies. [15] *Acinobacter lwoffi* K24, a soil bacterium, utilizes aniline as a source of nitrogen and carbon. It represents more than 20 aniline-cultured bacteria that produce 20 aniline-activated 2D-PAGE proteins, with the recognized protein spots present in the amino group transfer, malate dehydrogenase, the ketoadipate process, and HHDD isomerase. Biodegradable genes are reveale

GENOMICS IN BIOREMEDIATION

Genomics is an important computer innovation applied to understand the composition and functions of the genomic sequence of an organism that depends on analyzing the organism's complete DNA arrangement. The area involves thorough attempts to regulate the complete DNA sequence of organisms through in-depth genetic mapping attempts. The area also comprises knowledge of intragenomic theory like heterosis, pleiotropy, epistasis, and additional relations among loci and factors present inside the genome. In comparison, the examination of a single gene, its role, and functions, is common in present-times biological and medical research, and an important center of molecular biology, doesn't come under the description of genomic science until the goal of the genomic path and practical knowledge investigation is to explain its result, place in, feedback to the complete genome's network. To direct this huge extensive data, bioinformatics is of great use. Bioinformatics provides both theoretical bases and practical knowledge for finding structural and functional actions of the cell and the organism.

Table 1: List of bioinformatics branches and its application in bioremediation

Bioinformatics Tool/Technique	Role in Bioremediation
Sequence Analysis	Identification of pollutant-degrading genes and enzymes, phylogenetic analysis of microorganisms.
Genomics	Microbial genome sequencing to understand metabolic pathways and identification of potential bioremediation targets.
Metagenomics	Analysis of microbial communities in contaminated environments, discovery of novel biodegradation pathways.
Proteomics	Identification and characterization of proteins involved in biodegradation, understanding protein-protein interactions.
Metabolomics	Characterization of metabolic products and pathways, monitoring bioremediation progress.

Bioinformatics Tool/Technique	Role in Bioremediation
Machine Learning	Predictive modeling of biodegradation rates, and optimization of bioremediation processes.
Database and Information Systems	Storage and management of bioremediation data, knowledge sharing.
Structural Biology	Understanding the structure of biodegradation enzymes, designing enzyme variants for enhanced activity.

GENOMICS TOOLS AND TECHNIQUES APPLIED IN BIOREMEDIATION

The control of gene expression is one of the primary mechanisms for adjusting to changes in environmental conditions and ensuring survival. DNA microarrays are an incredibly dominant program that makes it possible to calculate the level of mRNA expression for each gene in a microorganism. The explanation of data is the most challenging aspect of microarray investigations. Different genes may be up- or down-regulated during a certain stress scenario. Despite having access to the complete genome sequence of microorganisms capable of bioremediation, research is not moving forward quickly.[16] Genome-wide DNA microarrays can be used to determine how each gene in the genome is expressed in a given situation. In the past, DNA microarrays were used to physically evaluate the culture medium and track the expression profiles of the catabolic genes in mixed microbial communities. More than 100 genes were discovered to be influenced by oxygen status when changes in mRNA expression levels in Bacillus subtilis growing under anaerobic conditions were examined using DNA microarrays. Frailty is typically a problem because PCR-based cDNA microarrays only identify genes from people who contribute more than 5% of the DNA in the community. To validate the sensitivity of punctate oligonucleotide DNA microarrays and their use in functional genomics, several factors were examined. We discovered that the 40mM concentration of 5'-C6-amino-modified 70-mer CMT-GAPII substrates with the Triamide signal amplification tag had the most detrimental effects. For the best biodegradation populations, an enlarged 50- -mer oligonucleotide microarray based on the most frequently used genes and pathways involved in biodegradation and metal resistance was constructed. This type of DNA microarray was successfully used to study the enrichment of naphthalene, and soil microcosms demonstrated how the microflora altered noticeably during incubation conditions. In addition, DNA microarrays are used in genome-wide transcriptional profiles, quantitative roles of stress gene research on microbial genomes, and regulation of specific bacterial species.

PHYLOGENY IN BIOREMEDIATION

The process of developing a theory about the evolutionary relationship of species based on distinguishing characteristics is known as phylogenetic interpretation. Traditionally, the focus of phylogenetic research has been on the analysis of all species. The earlier biologists grouped living things into the Symbolic Tree of Life before Linnaeus created the categorization method into kingdom, phylum, genus, and species. A suitable schematic for illustrating evolutionary links based on sequence similarity has been acquired, and it is called a "phylogenetic tree" since it is a tree that represents relationships between species. Because a parent branch can only break into two daughter branches at every branch point, phylogenetic trees are strictly binary. There are numerous methods for calculating the length of branches in a phylogenetic tree. One method of determining branch length is by comparing the evolution of sequences to other sequences in the input dataset. A phylogenetic tree is made up of sequence information that may or may not be rooted, whereas the phylogeny of a species often has a single root, indicating that the species had a common ancestor.

BIOINFORMATICS DATABASES APPLIED IN BIOREMEDIATION

1. KEGG- https://www.genome.jp/kegg/ Metabolic pathways, biological functions, and relationships.
2. BRENDA- https://en.wikipedia.org/wiki/Brenda Enzyme information, substrate specificity, and inhibitors.
3. UniProt- https://www.uniprot.org/ Protein sequences, functions, and interactions.
4. NCBI- https://www.ncbi.nlm.nih.gov/ GenBank, PubMed, biological databases, and taxonomy.
5. EnzyBase- https://bmcmicrobiol.biomedcentral.com/articles/10.1186/14 71-2180-12-54 Structures, roles, and kinetics of enzymes.
6. MetaCyc- https://metacyc.org/ Enzymes and metabolic processes in different species.
7. BioCyc- https://biocyc.org/ Databases of pathways and genomes for diverse creatures.
8. SEED- https://www.seedheritage.com/ A genome-based functional annotation and categorization system.
9. PATRIC- https://ngdc.cncb.ac.cn/databasecommons/database/id/230 Pathogenic microbes' biology and genomics.
10. IMG\JGI- https://img.jgi.doe.gov/ Metagenomes and microbial genomes.
11. MetaHIT- https://www.gutmicrobiotaforhealth.com/metahit/ Human gut microbiome metagenomic data.
12. MG-RAST- https://www.mg-rast.org/ Annotation and study of the metagenome.
13. magnify- https://www.ebi.ac.uk/metagenomics/ Interpretation and analysis of metagenomic data.
14. MicrobiomeDB- https://microbiomedb.org/ Tools for investigation of microbial genomes and metagenomes.
15. PICRUSt- https://picrust.github.io/picrust/ Predicts microbial communities' capacity for function using data from 16S rRNA gene sequencing.

CONCLUSION

Many different features of a cell, including the use of genetic information and proteins as well as metabolic and regulatory pathways, have been described and studied using new bioinformatics methods. The study of technologies to comprehend the cellular systems that operate and govern germs will speed up and be made simpler by bioinformatics research. The next ten years will include the use of institutional bioinformatics to understand molecular processes and cellular regulation. Bioinformatics has many applications in the field of Bioremediation for the analysis of patterns and Biodegradation pathways of exogenous organic compounds.

REFERENCES

1. F. B. V. J. F. F. C. P. M. L. de Oliveira M, "Pharmaceuticals residues and xenobiotic contaminants:
2. Occurrence analytical techniques and sustainable alternatives for wastewater treatments," Sci Total Environ, 2020. C. DD, "Environmental Xenobiotics and their adverse health impacts- a general review," J Environ Poll Human Health, pp. 77-88, 2018.
3. B. A. B.-M. M. H. J. Zhu Y, "The toxic effects of xenobiotics on the health of humans and animals," Biomed Res Int, 2017.
4. S. R, "Biodegradation of Xenobiotics-a way for environmental detoxification," Int J Dev Res, pp. 14082-14087, 2017.
5. G. S. B. G. Z. W. M. D. M. S. Bhatt P, "New insights into the degradation of synthetic pollutants in contaminated environments," Chemosphere, 2020.
6. S. H. B. V. K. N. Mathew A, "Classification, source, and effect of environmental pollutants and their biodegradation," J Environ Pathol Toxicol Oncol, pp. 55-71, 2017. A. K. S. &. K. S. Kumar, "Bioremediation of heavy metals using some bioinformatics tools: A comprehensive review.," Journal of Environmental Chemical Engineering., pp. 4716-4727, 2018.
7. M. M. S. K. Chandran H, "Microbial Biodiversity and bioremediation assessment through omics approaches," Front Environ Chem, 2020.
8. S. S. P. Jaiswal, "Bioinformatics for microbial community analysis: Challenges and potential applications in bioremediation.," Journal of Environmental Management, 2020.
9. S. B. H. R. S. P. Dangi AK, "Bioremediation through microbes: systems biology and metabolic engineering approach," Crit Rev Biotechnol, pp. 79-98, 2018.
10. S. S. B. Y. A. J. K. J.,. L. L. Lee SY, "Proteomic analysis of polycyclic aromatic hydrocarbons (PAHs) degradation and detoxification in Sphingobium chungbukense DJ77," Microbiol Biotechnol, pp. 1943-1950, 2016.
B. M. N. M. K. M. R. M. Aslam B, "Proteomics: technologies and their applications," J Chromatogr Sci, pp. 182-196, 2017. C. P. C. C. Arsesne-Ploetze F, "Proteomic Tools to decipher microbial community structure and functioning," Environ Sci Pollut Res, pp. 13599-13612, 2015.
11. B. K. Q. C. M. C. A. &. C. C. D. Singh, " Bioinformatics insights into pollution-induced microbial shifts and biodegradation pathways.," Bioinformatics, pp. 2073-2084, 2018.
12. Y. W. J. &. T. M. Wu, "Bioinformatics tools and database for mining microbial degradation of pollutants.," Frontiers in Microbiology., 2019.
13. R. &. S. P. Singh, "Impact of bioinformatics on bioremediation: Computational tools and techniques for optimizing biodegradation," Current Research in Microbial Sciences, pp. 33-45, 2016.

CHAPTER 12

MICROBIAL GENOMICS AND ITS INDUSTRIAL APPLICATIONS

Abhishek Kumar Mishra and Ruchi Yadav

Amity Institute of Biotechnology, Amity University Uttar Pradesh, Gomti Nagar

Abstract
Microbial genomics, the study of the complete genetic material of microorganisms, has revolutionized various industries by offering powerful insights into microbial functions, diversity, and potential. By sequencing and analyzing microbial genomes, scientists can identify genes responsible for producing enzymes, metabolites, and bioactive compounds, enabling the development of novel industrial applications.In biotechnology, microbial genomics has enhanced the production of biofuels, bioplastics, and bio-based chemicals by optimizing microbial strains for higher efficiency and yield. In the pharmaceutical industry, the discovery of novel antibiotics, enzymes, and therapeutic compounds from microbial sources has been accelerated through genomic screening. Genomic data also supports the development of probiotics and microbiome-based therapies, which are increasingly used for health and wellness products. In agriculture, microbial genomics aids in the creation of biofertilizers and biopesticides, offering eco-friendly alternatives to chemical fertilizers and pesticides. Microbial genomics is transforming industrial practices by enabling the precise identification and utilization of microorganisms for diverse applications. As genomic technologies continue to advance, they will likely play an increasingly critical role in shaping the future of sustainable and innovative industrial processes.

Keywords:
Microbial genomics, genomics, microbial genetics, functional genomics

INTRODUCTION

Microbial genomics is the study of the genetic makeup of microorganisms, including bacteria, archaea, and viruses. It involves analyzing the DNA sequences of these organisms to understand their evolution, diversity, and functions within various environments and host organisms [1][2].

The significance of microbial genomics lies in its ability to provide insights into the complex interactions between microbes and their hosts, as well as the role of microbiota in health and disease. By studying the gut microbiome, researchers can identify changes in microbial composition (dysbiosis) associated with conditions like inflammatory bowel diseases (IBD), familial adenomatous polyposis (FAP), and colorectal cancer (CRC) [1][2].

Advances in next-generation sequencing technologies have enabled researchers to analyze the gut microbiome more comprehensively. By sequencing the 16S rRNA gene, which is specific to bacteria and archaea, researchers can identify the relative abundance of different microbial taxa in samples collected from FAP, UC, and healthy individuals [1][2].

Microbial genomics allows scientists to explore and catalog the immense diversity of microorganisms. Through genomic sequencing, the researcher can identify new species and understand their evolutionary relationship [3]. Genomic sequencing is a laboratory method that is used to determine the entire genetic makeup of a specific organism or cell type [4]. Genomics studies provide insights into the evolutionary mechanism that drives microorganisms' adaptation and survival in various environments. This helps in reconstructing the evolutionary history of life on Earth [5].

In Medical, Microbial Genomics enables the rapid identification and characterization of pathogenic microorganisms. This is crucial for diagnosing infectious diseases and developing effective treatments. Understanding the genetic basis of antibiotic resistance helps in developing strategies to combat resistant strains. Genomics also aids in the discovery of new antibiotics by revealing biosynthetic gene clusters [6].

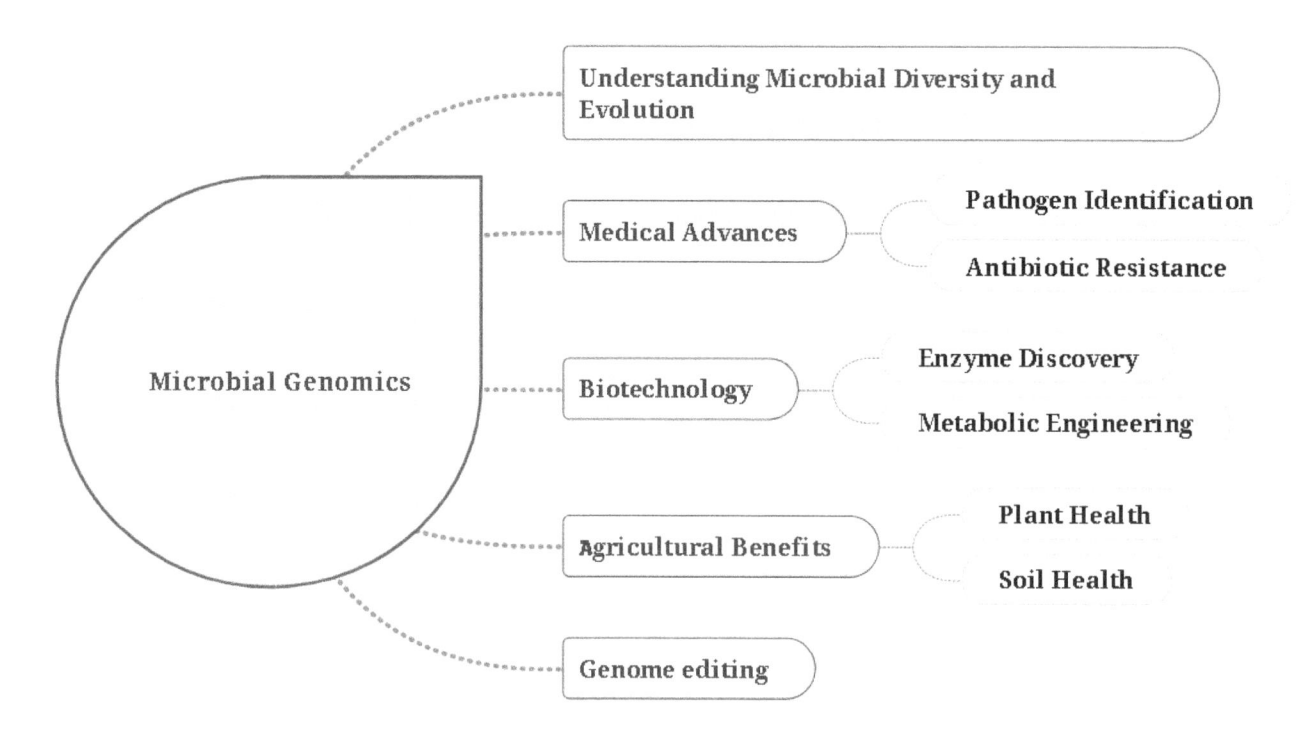

Figure 1: Significance of microbial genomics

In biotechnology, Microbial genomics facilitates the discovery of novel enzymes that can be used in various industrial processes, from food production to biofuel synthesis. By understanding microbial metabolic pathways, Scientists can engineer microorganisms to produce valuable compounds, such as pharmaceuticals, biofuels, and bioplastics [7].

In agriculture, Genomics reveals the interactions between plants and their associated microbes, leading to the development of biofertilizers and biopesticides that enhance plant growth and health. Studying the soil microbiome helps maintain soil health and fertility, which is essential for sustainable agriculture [8].

In Genome editing, Techniques like CRISPR-Cas9, informed by microbial genomics, allow precise editing of microbial genomes, opening new possibilities in synthetic biology for creating custom-designed microorganisms.

HISTORY AND DEVELOPMENT OF MICROBIAL GENOMICS

Waste Management Methods
BIOREMEDIATION OF INDUSTRIAL WASTE THROUGH ENZYME-PRODUCING MARINE MICROORGANISMS.

The bioremediation process using microorganisms is a kind of nature-friendly and cost-effective clean green technology. Recently, the biodegradation of industrial wastes using enzymes from marine microorganisms has been reported worldwide [12]. Bioremediation is a promising way of dealing with HM pollution. Microbes have the ability with various potencies to resist HMs tension [13]. The potentiality of microorganisms to degrade the pollutants is mainly attributed to their nutritional versatility and metabolic diversity, which can be exploited in the bioremediation process. Autochthonous microorganisms i.e., bacteria, fungi, yeast, and algae are capable of degrading and/or rendering these contaminants immobile in soil via adaptive mechanisms of biosorption, bioaccumulation, biotransformation, and biomineralization to aid their survivability in heavy metal contaminated soils [14]. Bioremediation can be a suitable alternative to the physicochemical approaches, which are environmentally destructive and costly and may cause secondary pollution [15]. Water pollution is also treated by bioremediation. Water pollution is on the rise because of increased human population and activities, unsustainable agricultural practices, and rapid industrialization, and it is a major global concern [16].

Industrial Waste Composting

Composting is an important technology used to treat and convert organic waste into value-added products. Recently, several studies have been done to investigate the effects of microbial supplementation on the composting of agro-industrial waste. The addition of microorganisms could enhance the composting process such as accelerating the organic matter degradation, mineralization, microbial enzyme activities, and the quality of the end-products [17]. Among the various waste management systems, composting is considered the sustainable treatment method for recycling agro-industrial waste into valuable by-products [18]. Composting is a self-heating aerobic biodegradative process of organic matter that is carried out under controlled conditions via microbial activities to produce valuable, stable, humus-rich products useful for the cultivation of crops and the soil. In addition, composting is an alternative to the landfill and incineration approaches for waste disposal. As a safe and economical process, several agriculture-based industries have adopted composting technology as a treatment method in their waste management system. The use of this system to convert organic waste into valuable decomposed by-products promotes the reduction of waste treatment costs. In general, industrial activities involving the processing of certain crops or livestock generate an abundance of agro-industrial residues. These residues present either in solid or liquid forms and are mainly composed of complex proteins, carbohydrates, and polyphenolic constituents [19,20]

Physicochemical Method

Coagulation is a well-known physicochemical method used to remove pollutants from textile wastewater. Alum and iron salts act as coagulants, helping small particles in the water combine into larger ones. Flocculation-coagulation processes are effective for removing dispersed dyes from wastewater, though they are less efficient at treating reactive and vat dyes.

Algae in Industrial Waste Management

Algae are an effective intermediary that can convert carbon dioxide and solar energy into various bio-energy forms (such as biodiesel, bio-ethanol, and bio-butanol) since they possess 20% higher photosynthetic efficiency compared to terrestrial plants. The development of an algae bacteria consortium system may be an ideal process to solve the above-mentioned obstacles associated with microalgae-based wastewater treatment process. In the co-cultivation system, the conventional activated sludge with bacteria can effectively remove the organic carbon source (COD), along with the generation of CO_2. Through photosynthesis, algae can convert CO_2 to biomass and produce O_2 to support the bacterial growth. The algae-bacteria consortium system can be perfectly applied for wastewater treatment to avoid the external oxygen supply, allow nutrient assimilation into biomass, and reduce CO_2 emissions to the atmosphere [22].

PHYTOREMEDIATION

Bioremediation and phytoremediation can be highly effective methods for addressing serious environmental pollution. Bioremediation involves using living organisms, such as bacteria, fungi, or plants, to break down or clean up harmful substances. Microorganisms can break down or transform pollutants, making them safe to release into the environment. Bioremediation has been utilized in diverse contexts, encompassing the mitigation of soil and water pollution caused by petroleum derivatives, heavy metals, pesticides, and organic substances [23,24,25,26,27]. Phytoremediation presents a potentially viable method for the remediation of textile industry effluent. Algae possess distinctive metabolic capabilities that facilitate the efficient elimination or degradation of pollutants, such as dyes [28,29].

Table 1: List of microorganism that are used for waste management

Microorganisms	Waste compounds
Pseudomonas putida	Benzene and xylene
Gloeophyllum trabeum, Trametes versicolor	Hydrocarbons
Acinetobacter sp., Micro bacterium sp.	Aromatic hydrocarbons
Pseudomonas cepacian, Bacillus cereus, Citrobacter koseri	Crude oil
Micrococcus luteus, Listeria denitrificans, Nocardia atlantica.	Textile dyes
Bacillus, Staphylococcus	Endosulfan

CONCLUSION

Industrial microbiology and waste management are deeply intertwined fields that play a crucial role in advancing sustainable practices and environmental protection. Effective industrial management is crucial for optimizing productivity, ensuring sustainability, and maintaining competitiveness in today's dynamic market. By implementing best practices in leadership, technology, and operations, organizations can streamline processes, enhance employee performance, and foster innovation. Prioritizing strategic planning, embracing digital transformation, and cultivating a culture of continuous improvement is essential for achieving long-term success. Adopting these principles not only drives efficiency and profitability but also positions companies to adapt to future challenges and opportunities in the industrial landscape.

REFERENCES

1. Abdel-Mohsen O. Mohamed, M El Gamal, Suhaib M. Hameedi, Evan K. (2023) Paleologos, Chapter 1 - Emerging carbon-based waste management sustainable practices, Editor(s): Abdel-Mohsen O. Mohamed, M El Gamal, Suhaib M. Hameedi, Evan K. Paleologos, Sustainable Utilization of Carbon Dioxide in Waste Management, Elsevier, Pages 1-66,
2. Adebayo, F.O., Obiekezie, S.O.. Microorganisms in Waste Management. Research J. Science and Tech. 2018; 10(1): 28-39 doi: 10.5958/2349-2988.2018.00005.0.
3. Adebayo, Fatimat & Smart, Obiekezie. (2018). Microorganisms in Waste Management. Research Journal of Science and Technology. 10. 10.5958/2349-2988.2018.00005.0.
4. Ezeonu, C. S., Tagbo, R., Anike, E. N., Oje, O. A. and Onwurah, I. N. E. (2012). Biotechnological Tools for Environmental Sustainability: Prospects and Challenges for Environments in Nigeria—A Standard Review. Biotechnology Research International. Vol 2012.
5. Capel C. Innovations in Waste. 2010. Available from URL: https://waste-management-world.com/a/1innovations-in-waste
6. TY - CHAPAU - Kareem, Hafiz Abdul, Riaz, Sobia, Sadia, Haleema, Mehmood, Rizwan-2022/08/05SN - 9781774910542 Industrial-Waste, Types, Sources, Pollution Potential, Country Wise ComparisonsDO
7. Nahiun, Khandakar & Sarker, Bijoyee & Keya, Kamrun N. & Mahir, Fatin & Shahida, Shahirin & Khan, Ruhul. (2021). *Scientific Review A Review on the Methods of Industrial Waste Water Treatment. Scientific Review.* 7. 20-31. 10.32861/sr.73.20.31 In Industrial Recycling, Recycling, Waste Management by David FahrionJuly 11, 2019
8. Adebayo, F.O., Obiekezie, S.O.(2018) Microorganisms in Waste Management. Research J. Science and Tech. 2018; 10(1): 28-39
9. Awaleh, Mohamed & Soubaneh, Youssouf. (2014). Waste Water Treatment in Chemical Industries: The Concept and Current Technologies. Hydrology Current Research
10. Adebayo, Fatimat & Smart, Obiekezie. (2018) Microorganisms in Waste Management. Research Journal of Science and Technology. 10.5958/23492988.2018.00005.0
11. P. Sivaperumal, K. Kamala, R. Rajaram, Chapter Eight - Bioremediation of Industrial Waste Through Enzyme Producing Marine Microorganisms, Editor(s): Se-Kwon Kim, Fidel Toldrá, Advances in Food and Nutrition Research, Academic Press, Volume 80,2017, Pages 165-179, ISSN 1043-4526, ISBN 9780128095874,
12. Alabssawy, A.N., Hashem, A.H. Bioremediation of hazardous heavy metals by marine microorganisms: a recent review. Arch Microbiol 206, 103 (2024).

13. Dr. Javid A. Parray, Dr. Abeer Hashem Abd Elkhalek Mahmoud, Prof. Riyaz Sayyed First published: 19 March 2021.
14. Bhargava, R.N., Saxena, G., Mulla, S.I. (2020). Introduction to Industrial Wastes Containing Organic and Inorganic Pollutants and Bioremediation Approaches for Environmental Management. In: Saxena,
15. G., Bhargava, R. (eds) Bioremediation of Industrial Waste for Environmental Safety. Springer, Singapore. https://doi.org/10.1007/978-981-13-1891-7_1
16. Nishita Ojha et al 2021 IOP Conf. Ser.: Earth Environ. Sci. 796 012012DOI 10.1088/17551315/796/1/012012 Submission received: 16 December 2021 / Revised: 9 January 2022 / Accepted: 10 January 2022 / Published: 14 January 2022
17. Lim, S.L.; Lee, L.H.; Wu, T.Y. Sustainability of using composting and vermicomposting technologies for organic solid waste biotransformation: Recent overview, greenhouse gases emissions, and economic analysis. J. Clean. Prod. 2016, 111, 262–278.
18. Vallejos, M.E.; Felissia, F.E.; Area, M.C. Hydrothermal treatments applied to agro- and forest-industrial waste to produce high added-value compounds. BioResources 2016, 12, 2058Yusuf, M. Agroindustrial
19. waste materials and their recycled value-added applications: Review. In Handbook of Ecomaterials; Springer International Publishing: New York, NY, USA, 2017; pp. 1–11. ISBN 9783319482811.
20. Holkar C. R., Jadhav A. J., Pinjari D. V., Mahamuni N. M., and Pandit A. B., A critical review on textile wastewater treatments: possible approaches, Journal of Environmental Management. (2016) 182, 351–366, https://doi.org/10.1016/j.jenvman.2016.07.090, 2-s2.0-84982676403.
21. Yue Wang, Shih-Hsin Ho, Chieh-Lun Cheng, Wan-Qian Guo, Dillirani Nagarajan, Nan-Qi Ren, DuuJong Lee, Jo-Shu Chang, Perspectives on the feasibility of using microalgae for industrial wastewater treatment, Bioresource Technology, Volume 222, 2016, Pages 485-497, ISSN 0960-8524,
22. Masotti, F.; Garavaglia, B.S.; Gottig, N.; Ottado, J.Bioremediation of the Herbicide Glyphosate in Polluted Soils by Plant-Associated Microbes. Curr. Opin. Microbiol. 2023, 73, 102290.
23. Thacharodi, A.; Hassan, S.; Singh, T.; Mandal, R.; Chinnadurai, J.; Khan, H.A.; Hussain, M.A.; Brindhadevi, K.; Pugazhendhi, A. Bioremediation of Polycyclic Aromatic Hydrocarbons: An Updated Microbiological Review. Chemosphere 2023,328,138498.
24. Mohammed, S.A.; Omar, T.J.; Hasan, A.H. Degradation of Crude Oil and Pure Hydrocarbon Fractions by Some Wild Bacterial and Fungal Species. arXiv 2023, arXiv:2301.08715.
25. Ariyachandra, S.P.; Alwis, I.S.; Wimalasiri, E.M. Phytoremediation Potential of Heavy Metals by Cyperus Rotundus. Rev. Agric. Sci. 2023, 11, 20–35.
26. Kumar Gupta, P.; Ibrahim Mustapha, H.; Singh, B.; Chandra Sharma, Y. Sustainable Energy Technologies and Assessments Bioremediation of Petroleum Contaminated Soil-Water Resources Using Neat Biodiesel: A Review. Sustain. Energy Technol. Assess. 2022, 53, 102703.
27. Acuner, E.; Dilek, F.B. Treatment of Tectilon Yellow 2G by Chlorella vulgaris. Process Biochem. 2004, 39, 623–631.
28. Dubey, S.K.; Dubey, J.; Mehra, S.; Tiwari, P.; Bishwas, A.J. Potential Use of Cyanobacterial Species in Bioremediation of Industrial Effluents. Afr. J. Biotechnol. 2013, 10, 1125–1132.

CHAPTER **13**

GUT METAGENOMICS: A NEW AVENUE FOR DRUG DEVELOPMENT

Abhishek Kumar Mishra, Jyoti Prakash and Ruchi Yadav

Amity Institute of Biotechnology, Amity University Uttar Pradesh

Abstract

The human gut microbiome, composed of trillions of microorganisms, plays a vital role in health and disease. Metagenomics, the study of genetic material recovered directly from environmental samples, has emerged as a transformative tool in exploring the complex interactions between the gut microbiota and host physiology. In recent years, gut metagenomics has gained attention as a promising avenue for drug development. By providing insights into the functional capabilities of microbial communities, metagenomics enables the identification of novel therapeutic targets and biomarkers for various diseases, including gastrointestinal disorders, metabolic diseases, and neurodegenerative conditions Through the integration of sequencing technologies and bioinformatics, gut metagenomics offers a comprehensive understanding of microbial genes and their roles in drug metabolism, immunity modulation, and resistance mechanisms. This knowledge can drive the discovery of new drug candidates, probiotics, and personalized treatment strategies. Additionally, gut-derived metabolites, revealed through metagenomic analysis, have been linked to therapeutic potential, offering new approaches for drug repurposing and the development of microbiome-based therapies. This review highlights the key advancements in gut metagenomics and its implications for drug development.

Keywords:
Gut Metagenomics, Bioinformatics, Genomics, Drug design, Drug Development

INTRODUCTION TO GUT METAGENOMICS

Gut metagenomics is the comprehensive study of the genetic material within the gut microbiota. It involves analyzing the collective genome of all microorganisms present in the gastrointestinal tract, which includes bacteria, archaea, viruses, and eukaryotic microbes. The term "metagenomics" refers to the sequencing and analysis of DNA from environmental samples, allowing researchers to study the microbial communities without the need for culturing them in the lab [1].

The significance of gut metagenomics lies in its potential to unravel the complex interactions between the host and its microbiota. These interactions are crucial for various physiological processes, including digestion, immune function, and even behavior. By understanding the genetic diversity and functional capabilities of the gut microbiome, scientists can develop new strategies for treating diseases, improving gut health, and understanding overall human health better [1].

HISTORICAL CONTEXT AND ADVANCEMENTS

The concept of studying microbial communities in their natural environment's dates to the early 20th century, but significant advancements began with the advent of molecular biology techniques in the late 20th century [1]. Traditional microbiology methods, which relied on culturing microorganisms, were limited because many gut microbes are difficult or impossible to culture outside their natural habitat.

The introduction of DNA sequencing technologies in the 1970s, particularly the development of Sanger sequencing, allowed for the analysis of microbial genomes. However, it was the advent of next-generation sequencing (NGS) technologies in the early 21st century that revolutionized metagenomics. These high-throughput sequencing methods made it possible to sequence millions of DNA fragments simultaneously, providing a comprehensive view of microbial communities at an unprecedented depth and scale [1].

Key milestones in gut metagenomics include the Human Microbiome Project (HMP) launched in 2007, which aimed to characterize the human microbiome and its role in health and disease. This project significantly advanced our understanding of the gut microbiome and established a reference database for microbial communities in different body sites, including the gut [1].

OVERVIEW OF THE HUMAN GUT MICROBIOME

The human gut microbiome is a highly diverse and dynamic ecosystem comprising trillions of microorganisms. These microbes play a critical role in maintaining health by aiding in digestion, synthesizing vitamins, regulating the immune system, and protecting against pathogens [2][3][4]. The gut microbiome is dominated by bacteria, with Bacteroidetes and Firmicutes being the most abundant phyla. Other important groups include Actinobacteria, Proteobacteria, Verrucomicrobia, and Fusobacteria [2].

The composition of the gut microbiome varies between individuals and is influenced by factors such as diet, age, genetics, environment, and lifestyle [2][3]. It is established early in life, with significant changes occurring during infancy as the gut microbiota matures. By adulthood, the gut microbiome reaches a relatively stable state, though it can still be altered by diet, antibiotics, and other factors.

Research in gut metagenomics has revealed several key functions of the gut microbiome:
1. Digestion and Metabolism: Gut microbes break down complex carbohydrates, fibers, and proteins that the human digestive system cannot process alone. They ferment dietary fibers to produce short-chain fatty acids (SCFAs), which are important for gut health and energy metabolism [2,3].
2. Immune System Regulation: The gut microbiome influences the development and function of the immune system. It helps in the differentiation of immune cells and the production of antimicrobial compounds, maintaining the balance between immune tolerance and response [2,3].
3. Protection Against Pathogens: A healthy gut microbiome prevents colonization by harmful pathogens through competitive exclusion, production of antimicrobial substances, and modulation of the host's immune response [2,3,4].
4. Synthesis of Essential Nutrients: Certain gut bacteria synthesize essential vitamins, such as vitamin K and several B vitamins, contributing to the host's nutritional status [2,3].

Fig: Function of Gut Metagenomics

Understanding the gut microbiome through metagenomics offers insights into its role in various diseases, including inflammatory bowel disease (IBD), obesity, diabetes, and even mental health conditions. This knowledge paves the way for innovative treatments such as probiotics, prebiotics, and fecal microbiota transplantation (FMT), aimed at restoring or modifying the gut microbiota to improve health outcomes [2,3,4].

TECHNIQUES IN GUT METAGENOMICS

Sampling and Sequencing Methods

1. Sampling Methods:
Accurate and representative sampling is crucial in gut metagenomics to capture the diversity and functional capacity of the gut microbiota [5]. Common sampling methods include Fecal Samples the most widely used method due to its non-invasive nature and ability to reflect the composition of the gut microbiota. Fecal samples provide a good representation of the distal gut microbiome.

Biopsies obtained during colonoscopy or endoscopy, these samples can provide more localized information about specific regions of the gut but are more invasive and mucosal swabs these are less invasive than biopsies, these swabs can be used to sample the microbiota adherent to the gut mucosa, offering insights into microbial communities in close contact with the host tissues.

2. Sequencing Methods:

Once samples are collected, DNA is extracted and prepared for sequencing. Several sequencing methods are employed in gut metagenomics as mentioned below

- **16S rRNA Gene Sequencing:** This method targets the 16S ribosomal RNA gene, which is highly conserved among bacteria but contains hypervariable regions that allow for species identification. It is cost-effective and useful for taxonomic profiling but does not provide functional insights.
- **Whole Genome Sequencing (WGS):** Also known as shotgun metagenomics, this method sequences all the DNA present in a sample. It provides comprehensive data on the genetic content, including taxonomic and functional information. WGS is more informative but also more expensive and computationally demanding.
- **Meta transcriptomics:** This approach involves sequencing RNA to study the actively expressed genes in the microbiome, offering insights into microbial function and activity in real-time.
- **Metaproteomic and Metabolomics:** These methods analyze proteins and metabolites, respectively, to provide a deeper understanding of the functional state and metabolic capabilities of the gut microbiota.

BIOINFORMATICS TOOLS FOR DATA ANALYSIS

The large and complex datasets generated by metagenomic studies require robust bioinformatics tools for analysis. This table categorizes the tools based on their role in the metagenomic workflow, providing a brief description of each tool's function [5]. Key steps and tools include:

Category	Tool	Description
Quality Control	Trimmomatic	Trims low-quality bases and removes sequencing adapters from raw reads.
Pre-processing	Cutadapt	Removes adapter sequences and trims low-quality bases from sequencing reads.
Taxonomic Profiling	QIIME	A pipeline for analyzing 16S rRNA gene sequencing data for taxonomic classification and diversity analysis.
Tazonomic visualization	Kraken and Karona	Assigns taxonomic labels to metagenomic sequences with high accuracy and efficiency.
Functional Profiling	HUMAnN	Profiles the functional potential of microbial communities from metagenomic and metatranscriptomic data.
Functional Profiling	MetaPhlAn	Uses clade-specific marker genes for taxonomic profiling, useful in conjunction with HUMAnN.
Assembly and Annotation	MEGAHIT	An assembler used for reconstructing microbial genomes from metagenomic data.
Assembly and Annotation	SPAdes	Another assembler for metagenomic data, often used in microbial genome assembly.
Assembly and Annotation	Prokka	Annotates microbial genomes, providing insights into gene functions and metabolic pathways.
Statistical and Comparative Analysis	LEfSe	Identifies biomarkers and compares microbial communities across different conditions or groups.
Statistical and Comparative Analysis	STAMP	Facilitates statistical comparison of metagenomic profiles and visualizing differences in microbial communities.

CHALLENGES AND LIMITATIONS

Despite the advancements in gut metagenomics, several challenges and limitations persist:
1. Sampling Bias: The method of sample collection, storage, and processing can introduce biases. Fecal samples might not accurately represent the entire gut microbiota, particularly those associated with mucosal surfaces.
2. Complexity and Heterogeneity: The gut microbiome is highly diverse and varies between individuals and over time. Capturing this dynamic complexity requires large sample sizes and longitudinal studies, which can be resource intensive.
3. Data Analysis and Interpretation: The vast amount of data generated requires significant computational resources and expertise in bioinformatics. Interpreting functional data from metagenomes remains challenging due to the incomplete annotation of many microbial genes.
4. Contamination and Noise: Contaminants from the environment, reagents, or human DNA can affect the accuracy of metagenomic analyses. Rigorous quality control and contamination mitigation strategies are essential.
5. Cost and Accessibility: High-throughput sequencing and advanced computational analysis can be expensive, limiting accessibility for some researchers and institutions, particularly in low-resource settings.
6. Functional Insights: While taxonomic profiling is relatively straightforward, linking microbial taxa to specific functions and understanding their interactions with the host remains complex. Multi-omics approaches combining metagenomics with meta transcriptomics, metaproteomic, and metabolomics are needed to gain comprehensive functional insights.

GUT METAGENOME ROLE IN DIGESTION AND METABOLISM

The gut microbiome plays a pivotal role in human digestion and metabolism. The trillions of microorganisms residing in the gastrointestinal tract assist in breaking down complex carbohydrates, proteins, and fibers that the human digestive system cannot process on its own. Figure 1 shows the role of gut microbiome in digestion and metabolism. These microbes ferment indigestible fibers to produce short-chain fatty acids (SCFAs) like acetate, propionate, and butyrate, which are vital for colon health and serve as energy sources for colonocytes. Additionally, the gut microbiota synthesizes essential vitamins, including vitamin K and several B vitamins, which contribute to overall nutritional status. By influencing the gut's metabolic environment, the microbiome helps regulate energy balance and lipid metabolism, playing a significant role in weight management and metabolic health.[6]

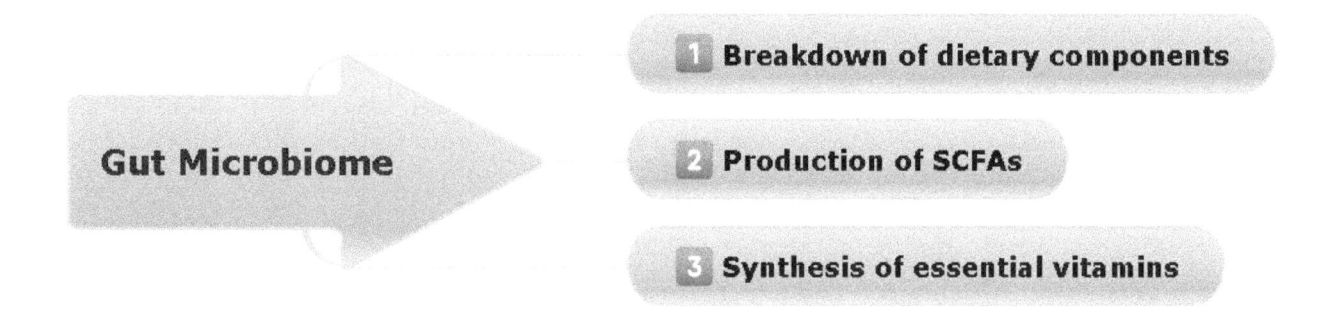

Figure 1: Role of Gut Microbiome in Digestion and Metabolism

GUT METAGENOME AND IMMUNE SYSTEM

The gut microbiome significantly influences the immune system, helping to shape its development and function. The interaction between gut microbes and the immune system begins early in life and continues to evolve, with the microbiome training the immune system to differentiate between harmful pathogens and benign antigens. Gut microbes contribute to the production of antimicrobial peptides and modulate the immune response, ensuring a balanced state of immune tolerance and defense. For example, certain bacterial species stimulate the production of regulatory T cells (Tregs), which help maintain immune homeostasis and prevent inflammatory responses [7]. This complex interplay is crucial for protecting the host from infections and inflammatory diseases.

DISEASES AND DISORDERS

The composition and function of the gut microbiome are closely linked to various diseases and disorders. Changes in the microbiome, often referred to as dysbiosis, can contribute to the development and progression of several health conditions.

Inflammatory Bowel Disease (IBD) Dysbiosis is strongly associated with IBD, including Crohn's disease and ulcerative colitis. Patients with IBD often exhibit reduced microbial diversity and an imbalance of beneficial and harmful bacteria, leading to chronic inflammation in the gut [8].

The gut microbiome influences energy harvest from the diet and fat storage. Certain microbial profiles are associated with increased energy extraction from food, contributing to obesity. Studies have shown that the microbiota of obese individuals can differ significantly from that of lean individuals, with a higher proportion of Firmicutes to Bacteroidetes [9,10]. Gut microbiome affects glucose metabolism and insulin sensitivity. Dysbiosis can lead to impaired glucose tolerance and insulin resistance, contributing to the development of type 2 diabetes. Specific microbial metabolites, such as SCFAs, play a role in modulating insulin signaling pathways [8].

DRUG DISCOVERY THROUGH GUT METAGENOMICS

Identification of Novel Drug Targets

Gut metagenomics has revolutionized the identification of novel drug targets by revealing the extensive genetic and functional diversity of the gut microbiota. This process involves several steps:
1. Metagenomic Sequencing and Annotation: Comprehensive sequencing of the gut microbiome provides a catalog of genes and pathways present in the microbial community. Bioinformatics tools like Prokka and HUMAnN help annotate these genes and predict their functions, highlighting potential targets related to metabolic pathways, enzyme activities, and signaling mechanisms. [11]

2. Functional Screening: Functional metagenomics involves expressing metagenomic DNA in heterologous hosts to screen for phenotypes of interest, such as antibiotic resistance or metabolite production. High-throughput screening techniques can identify genes and enzymes with desirable activities that could be targeted or utilized for therapeutic purposes.
3. Pathway Analysis: Detailed analysis of metabolic pathways and microbial interactions can pinpoint specific microbial functions that influence host health. Tools like MetaCyc and KEGG databases aid in mapping these pathways and identifying key nodes that could serve as drug targets. [11]
4. Host-Microbe Interaction Studies: Understanding the molecular dialogue between host and microbiota is crucial. Techniques such as transcriptomics and proteomics can reveal host responses to microbial signals and identify host targets modulated by microbial metabolites. Studies of host immune responses to microbiota can uncover immune-modulating targets for drug development.

SCREENING FOR NEW BIOACTIVE COMPOUNDS

The gut microbiome is a prolific source of bioactive compounds, including antibiotics, anti-inflammatory agents, and metabolites with therapeutic potential. The process of screening for these compounds involves:

- Culture-Independent Techniques: Metagenomic libraries are constructed from gut microbiome samples and screened for bioactivity in heterologous hosts or cell-based assays. High-throughput sequencing and bioinformatics analysis help identify genes responsible to produce bioactive compounds.[12]
- Metabolomic profiling of gut microbiota involves the comprehensive analysis of small molecules produced by microbes. Tools like Mass Spectrometry (MS) and Nuclear Magnetic Resonance (NMR) spectroscopy are used to identify and quantify these metabolites. Databases like HMDB (Human Metabolome Database) assist in metabolite identification and linking them to microbial origins. [13]
- Synthetic biology techniques enable the engineering of microbial strains to enhance the production of bioactive compounds. Gene clusters identified through metagenomics can be transferred into model organisms like E. coli for optimized production and further characterization. [12]
- Bioactive fractions from gut microbiome extracts are separated and tested for specific activities, such as antimicrobial or anti-inflammatory effects. This process helps isolate and identify novel compounds with therapeutic potential [14]

CASE STUDY

Several successful discoveries have emerged from gut metagenomics, highlighting its potential in drug discovery figure 2 shows the flow chart for drug discovery through Gut Metagenomics and table 2 list the key tools in gut metagenomics for drug discovery

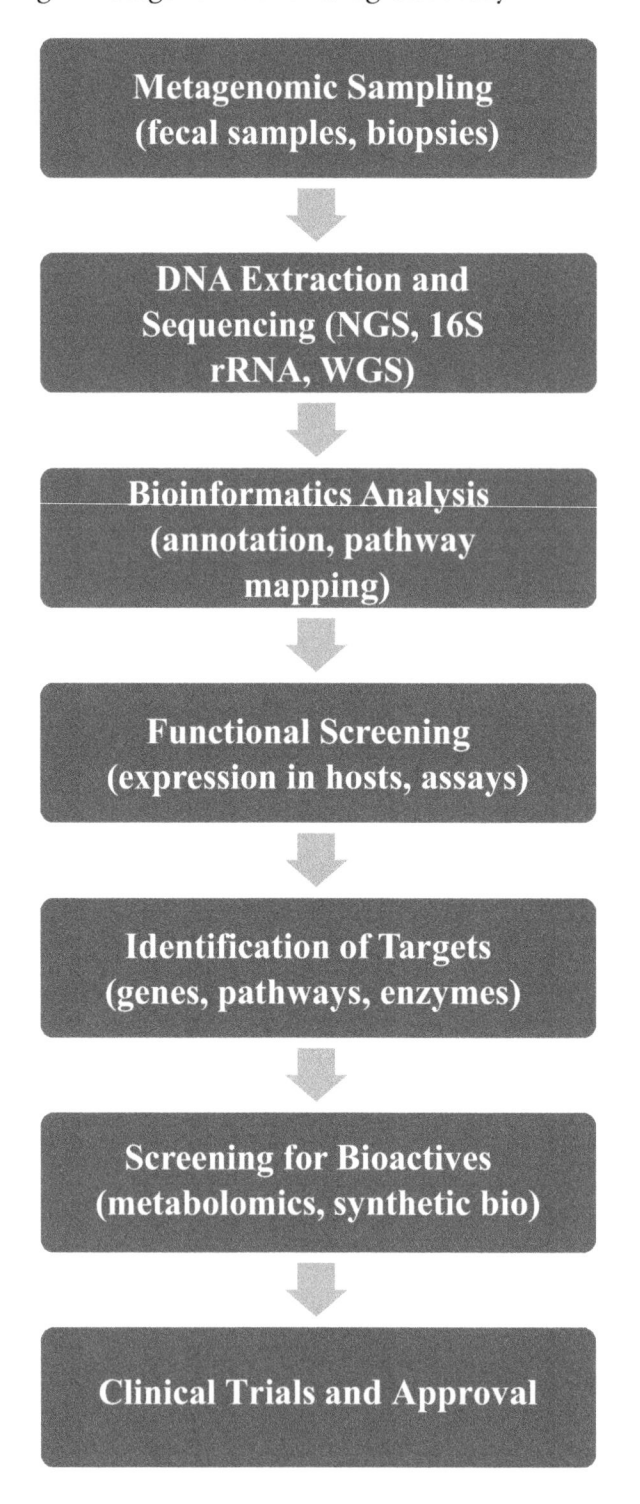

Figure 2: Steps for Drug Discovery through Gut Metagenomics

Table 1: Key tools in gut metagenomics for drug discovery

Step	Tool/Technique	Description
Sampling	Fecal Sampling, Biopsies	Methods to collect gut microbiota samples from individuals.
DNA Extraction & Sequencing	NGS, 16S rRNA, WGS	High-throughput sequencing techniques to analyze microbial genomes and diversity.
Bioinformatics Analysis	Prokka, HUMAnN, MetaPhlAn	Tools for annotating genes, profiling functional pathways, and taxonomic classification.
Functional Screening	Expression in Hosts	Screening metagenomic libraries for phenotypes of interest using heterologous expression systems.
Metabolomics	MS, NMR	Techniques for identifying and quantifying microbial metabolites.
Synthetic Biology	Gene Cloning, Expression	Engineering microbial strains for enhanced production of bioactive compounds.
Activity-Guided Fractionation	Fractionation Techniques	Separating bioactive compounds based on their activities for further identification and characterization.
Clinical Development	Drug Testing, Trials	Optimization of bioactive compounds into therapeutic drugs and conducting clinical trials for safety and efficacy.

Antibiotics like Lugdunin was discovered from the nasal microbiota, this antibiotic shows potent activity against methicillin-resistant Staphylococcus aureus (MRSA). The discovery was facilitated by metagenomic analysis and subsequent functional assays.

Teixobacti was identified through a novel culturing method (iChip), this antibiotic target lipid II, a precursor for cell wall synthesis, and is effective against various Gram-positive pathogens. [14]

Anti-inflammatory Compounds like Faecalibacterium prausnitzii this gut bacterium produces anti-inflammatory metabolites, including butyrate, which have shown potential in treating inflammatory bowel diseases (IBD). Indole propionic Acid: A microbial metabolite that exhibits neuroprotective properties and has been implicated in reducing the risk of diabetes.

Probiotics and Prebiotics example Akkermansia muciniphila was identified as beneficial for metabolic health, this bacterium has been developed into a probiotic formulation showing promise in improving insulin sensitivity and reducing inflammation. Human Milk Oligosaccharides (HMOs): Prebiotics derived from gut metagenomic studies have led to the development of HMOs that promote the growth of beneficial bacteria like Bifidobacteria in infants [14]

FUTURE DIRECTIONS AND INNOVATIONS IN GUT METAGENOMICS

The future of gut metagenomics lies significantly in personalized medicine and microbiome-based therapies. The understanding that everyone's microbiome is unique and has a profound impact on health opens the door for highly personalized treatments. Personalized medicine in this context involves tailoring medical treatment to the individual characteristics of each patient's microbiome. For example, by analyzing an individual's gut microbiome, healthcare providers can develop customized probiotic and prebiotic treatments aimed at modulating the gut flora to improve health outcomes [15,16].

Microbiome-based therapies are gaining traction, especially in the treatment of conditions like inflammatory bowel disease (IBD), irritable bowel syndrome (IBS), obesity, diabetes, and even mental health disorders such as depression and anxiety. Fecal microbiota transplantation (FMT) is one such therapy where stool from a healthy donor is transplanted into the gastrointestinal tract of a patient. This has shown promising results in treating recurrent Clostridium difficile infections and is being explored for other conditions.

The development of microbiome modulators, such as engineered probiotics, aims to introduce or enhance specific microbial functions within the gut. These probiotics can be designed to produce therapeutic compounds, degrade harmful substances, or stimulate the immune system [15]. The potential of microbiome-based therapies in personalized medicine is vast, promising more effective and less invasive treatments tailored to the unique microbial composition of everyone.

The future of gut metagenomics will be heavily influenced by advances in various technologies, particularly CRISPR and synthetic biology. CRISPR Technology: - CRISPR (Clustered Regularly Interspaced Short Palindromic Repeats) technology has revolutionized genetic engineering by allowing precise edits to DNA. In gut metagenomics, CRISPR can be used to modify the genomes of gut bacteria, enhancing their beneficial properties or eliminating harmful ones. This precise manipulation of microbial genes could lead to the development of more effective probiotics or targeted therapies for gut-related diseases [16]. Synthetic biology involves redesigning organisms for useful purposes by engineering them to have new abilities. In the context of gut metagenomics, synthetic biology can be used to create engineered microbes with specific functions, such as producing therapeutic molecules, detecting, and responding to disease markers, or enhancing nutrient absorption [16].

CONCLUSION

Gut metagenomics stands at the forefront of biomedical research, offering unprecedented insights into the complex and dynamic ecosystem of the human gut microbiome. By analyzing the collective genetic material of gut microorganisms, scientists can unravel the intricate interactions between the host and its microbiota, which are crucial for various physiological processes, including digestion, immune regulation, and overall health.

The advancements in sequencing technologies and bioinformatics tools have propelled the field of gut metagenomics, enabling comprehensive characterization of microbial communities and their functional capabilities. This progress has facilitated the identification of novel drug targets, the discovery of new bioactive compounds, and the development of innovative therapies aimed at modulating the gut microbiome to improve health outcomes.

The potential of gut metagenomics in drug discovery is immense, with successful examples already emerging, such as the development of probiotics, prebiotics, and antibiotics derived from microbial metabolites. Additionally, microbiome-based therapies, including fecal microbiota transplantation (FMT) and personalized medicine approaches, are gaining traction and showing promise in treating a range of conditions, from inflammatory bowel disease to metabolic and mental health disorders.

As the field continues to evolve, future directions in gut metagenomics will be driven by technological advancements, particularly in CRISPR and synthetic biology, which will enable precise manipulation and engineering of microbial communities. However, ethical and regulatory considerations must be addressed to ensure the safe, equitable, and responsible application of these technologies.

REFERENCES

1. Alessio Pini Prato, Bartow-McKenney, C., Hudspeth, K., Mosconi, M., Rossi, V., Stefano Avanzini, ... Cavalieri, D. (2019). A Metagenomics Study on Hirschsprung's Disease Associated Enterocolitis: Biodiversity and Gut Microbial Homeostasis Depend on Resection Length and Patient's Clinical History. *Frontiers in Pediatrics*, 7. https://doi.org/10.3389/fped.2019.00326

2. Seal, C. J., Courtin, C. M., Venema, K., & Jan de Vries. (2021). Health benefits of whole grain: effects on dietary carbohydrate quality, the gut microbiome, and consequences of processing. *Comprehensive Reviews in Food Science and Food Safety*, 20(3), 2742–2768. https://doi.org/10.1111/1541-4337.12728

3. Marie-Luise Puhlmann, & Willem. (2022). Intrinsic dietary fibers and the gut microbiome: Rediscovering the benefits of the plant cell matrix for human health. *Frontiers in Immunology*, 13. https://doi.org/10.3389/fimmu.2022.954845

4. Leeuwendaal, N. K., Stanton, C., O'Toole, P. W., & Beresford, T. P. (2022). Fermented Foods, Health and the Gut Microbiome. *Nutrients*, 14(7), 1527–1527. https://doi.org/10.3390/nu14071527

5. Sehli, S., Imane Allali, Rajaa Chahboune, Bakri, Y., Najib Al Idrissi, Hamdi, S., ... Ghazal, H. (2021). Metagenomics Approaches to Investigate the Gut Microbiome of COVID-19 Patients. *Bioinformatics and Biology Insights*, 15, 117793222199942-117793222199942. https://doi.org/10.1177/1177932221999428

6. Noronha, A., Modamio, J., Jarosz, Y., Guerard, E., Sompairac, N., Preciat, G., ... Puente, A. (2018). The Virtual Metabolic Human database: integrating human and gut microbiome metabolism with nutrition and disease. *Nucleic Acids Research*, 47(D1), D614–D624. https://doi.org/10.1093/nar/gky992

7. Carucci, L., Coppola, S., Luzzetti, A., Giglio, V., Vanderhoof, J., & Roberto Berni Canani. (2021). The role of probiotics and postbiotics in modulating the gut microbiome-immune system axis in the pediatric age. *Minerva Pediatrics*, 73(2). https://doi.org/10.23736/s2724-5276.21.06188-0

8. Md. Rayhan Mahmud, Akter, S., Sanjida Khanam Tamanna, Mazumder, L., Israt Zahan Esti, Banerjee, S., ... Anna Maria Pirttilä. (2022). Impact of gut microbiome on skin health: gut-skin axis observed through the lenses of therapeutics and skin diseases. *Gut Microbes*, 14(1). https://doi.org/10.1080/19490976.2022.2096995

9. Lim, J.-M., Vengadesh Letchumanan, Loh Teng-Hern Tan, Hong, K.-W., Wong, S.-H., Nurul-Syakima Ab Mutalib, ... Jodi Woan-Fei Law. (2022). Ketogenic Diet: A Dietary Intervention via Gut Microbiome Modulation for the Treatment of Neurological and Nutritional Disorders (a Narrative Review). *Nutrients*, 14(17), 3566–3566. https://doi.org/10.3390/nu14173566

10. Bozomitu, L., Miron, I., Anca Adam Raileanu, Lupu, A., Paduraru, G., Florin Mihai Marcu, ... Vasile Valeriu Lupu. (2022). The Gut Microbiome and Its Implication in the Mucosal Digestive Disorders. *Biomedicines*, 10(12), 3117–3117. https://doi.org/10.3390/biomedicines10123117

11. Beresford-Jones, B. S., Forster, S., Stares, M. D., Notley, G., Viciani, E., Browne, H., ... Pedicord, V. A. (2021). Functional and taxonomic comparison of mouse and human gut microbiotas using extensive culturing and metagenomics. Retrieved June 2, 2024, from bioRxiv website: https://www.semanticscholar.org/paper/Functional-and-taxonomic-comparison-of-mouse-and-Beresford-Jones-Forster/1e6224d3444764439d6b507f6700daa7e7e03c90

12. Zhang, X., J. Gangiredla, C. Tartera, Mammel, M., Barnaba, T., I. Edirisinghe, & B. Burton-Freeman. (2020). Gut Microbiome Metagenomics in Lean and Obese Individuals with Prediabetes and After Dietary Supplementation with Red Raspberry Fruit and Fermentable Fibers. Retrieved June 2, 2024, from Current Developments in Nutrition website: https://www.semanticscholar.org/paper/Gut-Microbiome-Metagenomics-in-Lean-and-Obese-with-Zhang-Gangiredla/a8b713aa0cb2baa308f57bcc7c9167a8918cf555
13. Song, C., Wang, B., Tan, J., Zhu, L., & Lou, D. (2017). Discovery of tauroursodeoxycholic acid biotransformation enzymes from the gut microbiome of black bears using metagenomics. *Scientific Reports*, 7(1). https://doi.org/10.1038/srep45495
14. Van, T., Hwangbo, H., Lai, Y., Seok Beom Hong, Choi, Y.-J., Park, H.-J., & Ban, K. (2023). The Gut-Heart Axis: Updated Review for The Roles of Microbiome in Cardiovascular Health. *Korean Circulation Journal*, 53(8), 499–499. https://doi.org/10.4070/kcj.2023.0048
15. Anthony, W. E., Burnham, C.-A. D., Dantas, G., & Kwon, J. H. (2020). The Gut Microbiome as a Reservoir for Antimicrobial Resistance. *the Journal of Infectious Diseases (Online. University of Chicago Press)/the Journal of Infectious Diseases*, 223(Supplement_3), S209–S213. https://doi.org/10.1093/infdis/jiaa497
16. Ju, Y., Wang, X., Wang, Y., Li, C., Yue, L., & Chen, F. (2022). [Application of metagenomic and culturomic technologies in fecal microbiota transplantation: a review]. *PubMed*, 38(10), 3594–3605. https://doi.org/10.13345/j.cjb.220573

CHAPTER **14**

INDUSTRIAL WASTE MANAGEMENT BY USING MICROORGANISMS

Priyanshi Dwivedi and Rachna Chaturvedi

Amity Institute of Biotechnology, Amity University Uttar Pradesh, Lucknow, 226028

Abstract

Microorganism plays a crucial role in various industries, including pharmaceuticals, agriculture, food and beverage, and environmental waste management. Waste management, on the other hand, deals with the collection, transportation, disposal, and recycling of waste materials. The intersection of industrial microbiology and waste management is particularly interesting because microorganisms can be harnessed to improve waste treatment processes and enhance environmental sustainability. Industrial waste management is a critical component of sustainable development, addressing the growing need to minimize environmental impacts and optimize resource use. This paper examines the current practices and challenges in industrial waste management, highlighting the importance of adopting innovative strategies for waste reduction, recycling, and reuse. With the rapid growth of industrial activities, the volume and complexity of waste have increased, necessitating the development of advanced technologies and management systems. Key strategies discussed include waste auditing, material flow analysis, and life cycle assessment, which help industries identify waste streams and opportunities for improvement. Effective industrial waste management requires robust policy and regulatory frameworks to ensure compliance and promote best practices. This review analyses the role of government regulations and international agreements in shaping waste management strategies across various industries.

Keywords:

Microorganism, industrial microbiology Waste management, Environmental sustainability.

INTRODUCTION

Industrial microbiology is a field of applied microbiology where microorganisms are used in various industrial processes. When discussing waste management, concepts like zero waste, sustainable materials management, circular economy, and others aim to reduce landfill waste by designing products that are both durable and environmentally friendly. Microorganisms play a key role in managing natural and man-made processes, helping make life easier. One significant area where they are used is waste management. Microorganisms involved in aerobic waste treatment include bacteria, fungi, algae, and protozoa, among others. The type of microorganisms that grow in a waste treatment system depends on the chemical properties of the waste, environmental factors, and the biological traits of the organisms. In aerobic systems, bacteria are the main organisms involved, as their biochemical diversity allows them to break down most organic compounds found in industrial waste. Both obligate aerobes and facultative bacteria are present in these systems. Microorganisms play an essential role in waste management, especially in dealing with solid waste, which can be divided into biodegradable (biowaste) and non-biodegradable categories. Biodegradable waste can be broken down by microorganisms and doesn't pollute for long periods. Anaerobic digestion, an innovative technology in waste management, is a recent and highly effective method for processing waste inside a closed system.

TYPES OF INDUSTRIAL WASTE

Industrial waste refers to any leftover material released from industrial activities, whether in gaseous, liquid, or solid form, regardless of whether the waste is categorized as industrial or domestic. This waste is often generated without proper disposal systems in place. Various types of waste produced by industries include cafeteria garbage, gravel, dirt, masonry, concrete, oil, chemicals, trash, scrap metals, solvents, lumber scrap, trees, weeds, solid and liquid waste, chemical waste, and toxic industrial effluents. Industrial waste can be classified into categories such as scrap lumber, gravel, scrap metal, solvents, masonry, oil, concrete, chemicals, and plastics. It may also include leftover food from restaurants, all of which can have a significant environmental impact. Water pollution is one of the most critical forms of pollution caused by industrial waste, along with pollution from other sources such as domestic waste, sewage, hazardous waste, municipal waste, medical waste, and manufacturing waste.

Industrial solid waste	Industrial liquid waste
Industrial chemical waste	**Industrial toxic waste**

Industrial waste

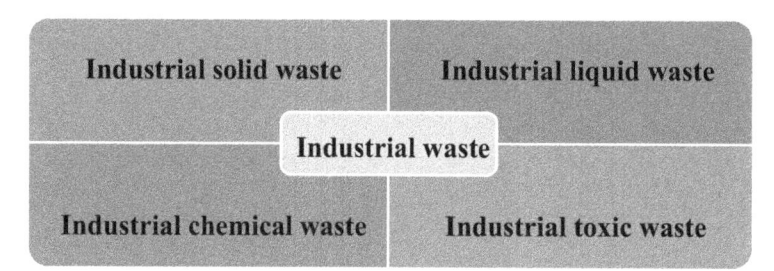

This figure represents types of industrial waste.

- **INDUSTRIAL LIQUID WASTE:** Liquid waste is generated from both households and industries. Many industrial processes require large amounts of water, which can become contaminated with hazardous substances such as radioactive materials, polluted water, organic liquids, rinse water, waste detergents, and even rainwater.

- **INDUSTRIAL SOLID WASTE:** Industrial solid waste consists of various materials such as paper, plastic, wood, cardboard, packaging, scrap metal, and other solid items that can no longer serve their original purpose.
- **INDUSTRIAL CHEMICAL WASTE:** Many industrial processes generate chemical waste, which includes flammable, corrosive, toxic, and explosive materials. Specialists should handle its disposal, as it often contains dangerous chemical residues that are harmful to humans. Animals and plant life.
- **INDUSTRIAL TOXIC WASTE:** Most chemical waste produced by industries, chemical plants, and garages is harmful and dangerous. If not correctly treated or disposed of, it can cause significant health and environmental problems. For this reason, regulations require that only government-authorized specialists handle its disposal. And specialized facilities [8].

MANAGEMENT OF INDUSTRIAL WASTE

Proper management of industrial waste prevents pollutants from contaminating air, water, and soil. This reduces the risk of harmful effects on ecosystems and wildlife. Microorganisms play an important role in industrial waste management systems it reviews the various roles of microorganisms in the environment, such as in sewage and soil treatment, energy generation, oil spillage, and radioactive contamination. It also discusses waste generation and management methods, and some specific uses of microorganisms (bacteria, fungi, algae, viruses, and protozoa) in waste management. It concludes by highlighting some recent advances in microbiological waste management.[9] Chemical industrial wastewater usually contains organic and inorganic matter in varying concentrations. Many materials in the chemical industry are toxic, mutagenic, carcinogenic, or simply almost non-biodegradable [10]. Microorganisms play important roles in the maintenance of many natural and man-made phenomena in the environment. They serve positive functions that make life easier and better for man. One such area where microorganisms are adopted is waste management. The various roles of microorganisms in the environment, such as in sewage and soil treatment, energy generation, oil spillage, and radioactive contamination. It also discusses waste generation and management methods, and some specific uses of microorganisms (bacteria, fungi, algae, viruses, and protozoa) in waste management [11].

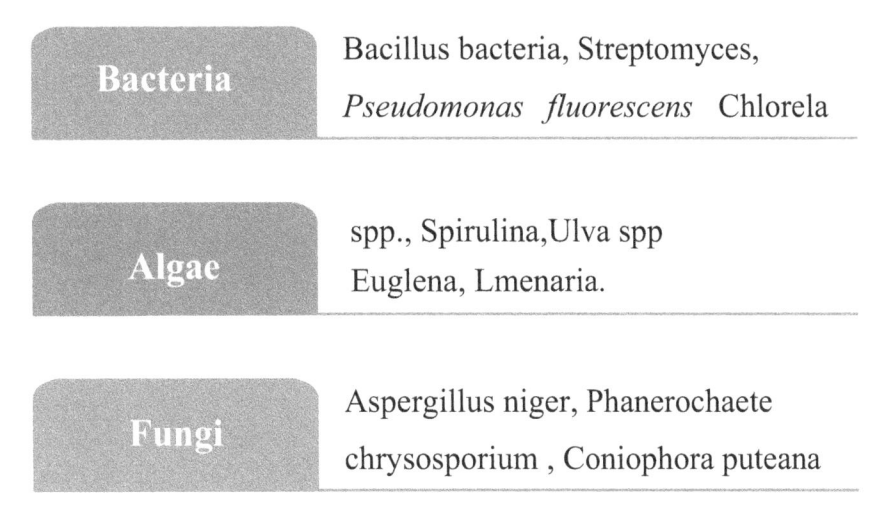

Bacteria	Bacillus bacteria, Streptomyces, *Pseudomonas fluorescens* Chlorela
Algae	spp., Spirulina, Ulva spp Euglena, Lmenaria.
Fungi	Aspergillus niger, Phanerochaete chrysosporium, Coniophora puteana

This figure represents microorganisms that are used for waste management.

WASTE MANAGEMENT METHODS

BIOREMEDIATION OF INDUSTRIAL WASTE THROUGH ENZYME-PRODUCING MARINE MICROORGANISMS.

The bioremediation process using microorganisms is a kind of nature-friendly and cost-effective clean green technology. Recently, the biodegradation of industrial wastes using enzymes from marine microorganisms has been reported worldwide [12]. Bioremediation is a promising way of dealing with HM pollution. Microbes have the ability with various potencies to resist HMs tension [13]. The potentiality of microorganisms to degrade the pollutants is mainly attributed to their nutritional versatility and metabolic diversity, which can be exploited in the bioremediation process. Autochthonous microorganisms i.e., bacteria, fungi, yeast, and algae are capable of degrading and/or rendering these contaminants immobile in soil via adaptive mechanisms of biosorption, bioaccumulation, biotransformation, and biomineralization in order to aid their survivability in heavy metal contaminated soils [14]. Bioremediation can be a suitable alternative to the physicochemical approaches, which are environmentally destructive and costly and may cause secondary pollution [15]. Water pollution is also treated by bioremediation. Water pollution is on the rise because of increased human population and activities, unsustainable agricultural practices, and rapid industrialization, and it is a major global concern [16].

INDUSTRIAL WASTE COMPOSTING

Composting is an important technology used to treat and convert organic waste into value-added products. Recently, several studies have been done to investigate the effects of microbial supplementation on the composting of agro-industrial waste. The addition of microorganisms could enhance the composting process such as accelerating the organic matter degradation, mineralization, microbial enzyme activities, and the quality of the end-products [17]. Among the various waste management systems, composting is considered the sustainable treatment method for recycling agro-industrial waste into valuable by-products [18]. Composting is a self-heating aerobic biodegradative process of organic matter that is carried out under controlled conditions via microbial activities to produce valuable, stable, humus-rich products useful for the cultivation of crops and the soil. In addition, composting is an alternative to the landfill and incineration approaches for waste disposal. As a safe and economical process, several agriculture-based industries have adopted composting technology as a treatment method in their waste management system. The use of this system to convert organic waste into valuable decomposed by-products promotes the reduction of waste treatment costs. In general, industrial activities involving the processing of certain crops or livestock generate an abundance of agro-industrial residues. These residues present either in solid or liquid forms and are mainly composed of complex proteins, carbohydrates, and polyphenolic constituents [19,20]

PHYSICOCHEMICAL METHOD

Coagulation is a well-known physicochemical method used to remove pollutants from textile wastewater. Alum and iron salts act as coagulants, helping small particles in the water combine into larger ones. Flocculation-coagulation processes are effective for removing dispersed dyes from wastewater, though they are less efficient at treating reactive and vat dyes.

PHYCO BIOREMEDIATION

Algae are an effective intermediary that can convert carbon dioxide and solar energy into various bio-energy forms (such as biodiesel, bio-ethanol, and bio-butanol) since they possess 20% higher photosynthetic efficiency compared to terrestrial plants. The development of an algae bacteria consortium system may be an ideal process to solve the above-mentioned obstacles associated with microalgae-based wastewater treatment process. In the co-cultivation system, the conventional activated sludge with bacteria can effectively remove the organic carbon source (COD), along with the generation of CO_2. Through photosynthesis, algae can convert CO_2 to biomass and produce O_2 to support the bacterial growth. The algae-bacteria consortium system can be perfectly applied for wastewater treatment to avoid the external oxygen supply, allow nutrient assimilation into biomass, and reduce CO_2 emissions to the atmosphere [22]. **MYCOREMEDIATION**

Bioremediation and phytoremediation can be highly effective methods for addressing serious environmental pollution. Bioremediation involves using living organisms, such as bacteria, fungi, or plants, to break down or clean up harmful substances. Microorganisms can break down or transform pollutants, making them safe to release into the environment. Bioremediation has been utilized in diverse contexts, encompassing the mitigation of soil and water pollution caused by petroleum derivatives, heavy metals, pesticides, and organic substances [23,24,25,26,27]. Phytoremediation presents a potentially viable method for the remediation of textile industry effluent. Algae possess distinctive metabolic capabilities that facilitate the efficient elimination or degradation of pollutants, such as dyes [28,29].

A microorganism that is used for waste management

Microorganisms	Waste compounds
Pseudomonas putida	Benzene and xylene
Gloeophyllum trabeum, Trametes versicolor	Hydrocarbons
Acinetobacter sp., Micro bacterium sp.	Aromatic hydrocarbons
Pseudomonas cepacian, Bacillus cereus, Citrobacter koseri	Crude oil
Micrococcus luteus, Listeria denitrificans, Nocardia atlantica.	Textile dyes
Bacillus, Staphylococcus	Endosulfan

CONCLUSION

Industrial microbiology and waste management are deeply intertwined fields that play a crucial role in advancing sustainable practices and environmental protection. Effective industrial management is crucial for optimizing productivity, ensuring sustainability, and maintaining competitiveness in today's dynamic market. By implementing best practices in leadership, technology, and operations, organizations can streamline processes, enhance employee performance, and foster innovation. Prioritizing strategic planning, embracing digital transformation, and cultivating a culture of continuous improvement is essential for achieving long-term success. Adopting these principles not only drives efficiency and profitability but also positions companies to adapt to future challenges and opportunities in the industrial landscape.

REFERENCE:

1. Abdel-Mohsen O. Mohamed, M El Gamal, Suhaib M. Hameedi, Evan K. (2023) Paleologos, Chapter 1 - Emerging carbon-based waste management sustainable practices, Editor(s): Abdel-Mohsen O. Mohamed, M El Gamal, Suhaib M. Hameedi, Evan K. Paleologos, Sustainable Utilization of Carbon Dioxide in Waste Management, Elsevier, Pages 1-66,
2. Adebayo, F.O., Obiekezie, S.O.. Microorganisms in Waste Management. Research J. Science and Tech. 2018; 10(1): 28-39 doi: 10.5958/2349-2988.2018.00005.0.
3. Adebayo, Fatimat & Smart, Obiekezie. (2018). Microorganisms in Waste Management. Research Journal of Science and Technology. 10. 10.5958/2349-2988.2018.00005.0.
4. Ezeonu, C. S., Tagbo, R., Anike, E. N., Oje, O. A. and Onwurah, I. N. E. (2012). Biotechnological Tools for Environmental Sustainability: Prospects and Challenges for Environments in Nigeria—A Standard Review. Biotechnology Research International. Vol 2012.
5. Capel C. Innovations in Waste. 2010. Available from URL: https://waste-management-world.com/a/1innovations-in-waste
6. TY - CHAPAU - Kareem, Hafiz Abdul, Riaz, Sobia, Sadia, Haleema, Mehmood, Rizwan-2022/08/05SN - 9781774910542 Industrial-Waste, Types, Sources, Pollution Potential, Country Wise ComparisonsDO
7. Nahiun, Khandakar & Sarker, Bijoyee & Keya, Kamrun N. & Mahir, Fatin & Shahida, Shahirin & Khan, Ruhul. (2021). Scientific Review A Review on the Methods of Industrial Waste Water Treatment. Scientific Review. 7. 20-31.10.32861/sr.73.20.31
8. In Industrial Recycling, Recycling, Waste Management by David FahrionJuly 11, 2019
9. Adebayo, F.O., Obiekezie, S.O.(2018) Microorganisms in Waste Management. Research J. Science and Tech. 2018; 10(1): 28-39
10. Awaleh, Mohamed & Soubaneh, Youssouf. (2014). Waste Water Treatment in Chemical Industries: The Concept and Current Technologies. Hydrology Current Research
11. Adebayo, Fatimat & Smart, Obiekezie. (2018) Microorganisms in Waste Management. Research Journal of Science and Technology. 10.5958/23492988.2018.00005.0
12. P. Sivaperumal, K. Kamala, R. Rajaram, Chapter Eight - Bioremediation of Industrial Waste Through Enzyme Producing Marine Microorganisms, Editor(s): Se-Kwon Kim, Fidel Toldrá, Advances in Food and Nutrition Research, Academic Press, Volume 80,2017, Pages 165-179, ISSN 1043-4526, ISBN 9780128095874,
13. Alabssawy, A.N., Hashem, A.H. Bioremediation of hazardous heavy metals by marine microorganisms: a recent review. Arch Microbiol 206, 103 (2024).
14. Dr. Javid A. Parray, Dr. Abeer Hashem Abd Elkhalek Mahmoud, Prof. Riyaz Sayyed First published: 19 March 2021.
15. Bhargava, R.N., Saxena, G., Mulla, S.I. (2020). Introduction to Industrial Wastes Containing Organic and Inorganic Pollutants and Bioremediation Approaches for Environmental Management. In: Saxena, G., Bhargava, R. (eds) Bioremediation of Industrial Waste for Environmental Safety. Springer, Singapore.https://doi.org/10.1007/978-981-13-1891-7_1
16. Nishita Ojha et al 2021 IOP Conf. Ser.: Earth Environ. Sci. 796 012012DOI 10.1088/17551315/796/1/012012
17. Submission received: 16 December 2021 / Revised: 9 January 2022 / Accepted: 10 January 2022 / Published: 14 January 2022

18. Lim, S.L.; Lee, L.H.; Wu, T.Y. Sustainability of using composting and vermicomposting technologies for organic solid waste biotransformation: Recent overview, greenhouse gases emissions, and economic analysis. J. Clean. Prod. 2016, 111, 262–278.
19. Vallejos, M.E.; Felissia, F.E.; Area, M.C. Hydrothermal treatments applied to agro- and forest-industrial waste to produce high added-value compounds. BioResources 2016, 12, 2058Yusuf, M. Agroindustrial
20. waste materials and their recycled value-added applications: Review. In Handbook of Ecomaterials; Springer International Publishing: New York, NY, USA, 2017; pp. 1–11. ISBN 9783319482811.
21. Holkar C. R., Jadhav A. J., Pinjari D. V., Mahamuni N. M., and Pandit A. B., A critical review on textile wastewater treatments: possible approaches, Journal of Environmental Management. (2016) 182, 351–366, https://doi.org/10.1016/j.jenvman.2016.07.090, 2-s2.0-84982676403.
22. Yue Wang, Shih-Hsin Ho, Chieh-Lun Cheng, Wan-Qian Guo, Dillirani Nagarajan, Nan-Qi Ren, DuuJong Lee, Jo-Shu Chang, Perspectives on the feasibility of using microalgae for industrial wastewater treatment, Bioresource Technology, Volume 222, 2016, Pages 485-497, ISSN 0960-8524,
23. Masotti, F.; Garavaglia, B.S.; Gottig, N.; Ottado, J.Bioremediation of the Herbicide Glyphosate in Polluted Soils by Plant-Associated Microbes. Curr. Opin. Microbiol. 2023, 73, 102290.
24. Thacharodi, A.; Hassan, S.; Singh, T.; Mandal, R.; Chinnadurai, J.; Khan, H.A.; Hussain, M.A.; Brindhadevi, K.; Pugazhendhi, A. Bioremediation of Polycyclic Aromatic Hydrocarbons: An Updated Microbiological Review. Chemosphere 2023,328,138498.
25. Mohammed, S.A.; Omar, T.J.; Hasan, A.H. Degradation of Crude Oil and Pure Hydrocarbon Fractions by Some Wild Bacterial and Fungal Species. arXiv 2023, arXiv:2301.08715.
26. Ariyachandra, S.P.; Alwis, I.S.; Wimalasiri, E.M. Phytoremediation Potential of Heavy Metals by Cyperus Rotundus. Rev. Agric. Sci. 2023, 11, 20–35.
27. Kumar Gupta, P.; Ibrahim Mustapha, H.; Singh, B.; Chandra Sharma, Y. Sustainable Energy Technologies and Assessments Bioremediation of Petroleum Contaminated Soil-Water Resources Using Neat Biodiesel: A Review. Sustain. Energy Technol. Assess. 2022, 53, 102703.
28. Acuner, E.; Dilek, F.B. Treatment of Tectilon Yellow 2G by Chlorella vulgaris. Process Biochem. 2004, 39, 623–631.
29. Dubey, S.K.; Dubey, J.; Mehra, S.; Tiwari, P.; Bishwas, A.J. Potential Use of Cyanobacterial Species in Bioremediation of Industrial Effluents. Afr. J. Biotechnol. 2013, 10, 1125–1132.

EPILOGUE

The journey through the "Industrial Perspective of Microbiology" has illustrated the profound impact that microorganisms have on various industries, ranging from pharmaceuticals and food production to environmental management and bioengineering. Throughout this exploration, we've seen how the application of microbiological principles and techniques has revolutionized traditional processes, leading to more efficient, sustainable, and innovative solutions.

In the pharmaceutical industry, microbiology has been pivotal in the development of antibiotics, vaccines, and biotechnology-based therapies. The ability to manipulate microbial genetics has opened new frontiers in personalized medicine and the fight against emerging infectious diseases.

The food and beverage industry continues to benefit from microbial fermentation processes, which not only enhance flavor and nutritional value but also ensure food safety and preservation. Probiotic research is also expanding, promising new health benefits through the manipulation of gut microbiota.

Environmental microbiology has shown us the potential for bioremediation and waste treatment, utilizing microorganisms to degrade pollutants and recycle waste materials. This field is crucial for developing sustainable practices that minimize environmental impact and promote ecological balance.

In industrial biotechnology, the use of microbes in biofuel production, bioplastics, and other bio-based materials is paving the way for a greener future. These innovations are reducing our reliance on fossil fuels and lowering carbon footprints, aligning with global efforts to combat climate change.

As we look to the future, the interdisciplinary nature of industrial microbiology will continue to foster collaboration across various scientific and engineering domains. The ongoing advancements in genomic technologies, synthetic biology, and bioinformatics will further enhance our ability to harness the power of microorganisms for industrial applications.

The promise of microbiology lies not only in its current applications but also in the untapped potential that future research will uncover. By continuing to invest in and prioritize microbiological research, we can anticipate groundbreaking developments that will address some of the most pressing challenges facing our world today.

In conclusion, the industrial perspective of microbiology is a testament to human ingenuity and the remarkable capabilities of microorganisms. As we advance our understanding and technological prowess, the integration of microbiology into industry will undoubtedly lead to a more sustainable, healthy, and prosperous future for all.